Planning by Design (PxD)–Based Systematic Methodologies

Planning by Design (PxD)–Based Systematic Methodologies

By

Hakan Bütüner

CRC Press
Taylor & Francis Group
Boca Raton London New York

CRC Press is an imprint of the
Taylor & Francis Group, an **informa** business

AN AUERBACH BOOK

CRC Press
Taylor & Francis Group
6000 Broken Sound Parkway NW, Suite 300
Boca Raton, FL 33487-2742

Fisrt issued in paperback 2020

ISBN-13: 978-1-4987-6132-1 (hbk)
ISBN-13: 978-0-367-65804-5 (pbk)

Library of Congress Cataloging-in-Publication Data

Names: Butuner, Hakan, author.
Title: Planning by design based systematic methodologies / Hakan Butuner.
Description: Boca Raton, FL : CRC Press, 2017. | Includes bibliographical references.
Identifiers: LCCN 2016047442 | ISBN 9781498761321 (hb : alk. paper)
Subjects: LCSH: Business planning. | Planning.
Classification: LCC HD30.28 .B879 2017 | DDC 658.4/012--dc23
LC record available at https://lccn.loc.gov/2016047442

Visit the Taylor & Francis Web site at
http://www.taylorandfrancis.com

and the CRC Press Web site at
http://www.crcpress.com

To my dear creative and contributive students

Contents

PART II SAMPLE SYSTEMATIC METHODOLOGIES

Preface

I felt the need to show how systematic planning can be applied to real-life cases even by professionals not skilled in Planning by Design. Accordingly, I wanted to write this book to enable Planning by Design (developed by the master, Richard "Dick" Muther, who has been involved in over 2000 projects in some 20 countries, has written 16 books with over 30 foreign-language translations, has taught worldwide in/for several well-known universities and organizations, and has founded four organizations …), this generalized masterguide, to be easily understood and to be universally applied to any type of subject area.

Richard Muther says: "Planning by Design is distinguished by following a pre-designed working model, a pre-designed working model that is compatible with the formula for planning, particular to the fundamentals of the subject area, and appropriate for the size of the project at hand."

Rather than explaining Planning by Design in detail, I preferred to introduce this book to show how it is used for developing working models in any type of subject area. The main headings in this book are as follows:

- The first section describes the nature of planning in general, the formula of planning, and the features that make it systematic. Additionally, the essence of Planning by Design is introduced.

- The second section demonstrates the application of creative systematic planning to real-life cases and the development of practical working models in different subject areas together with my dear students. Some of the working models are structured on the short form and some on the full version of the Planning by Design pattern; however, most of them are structured without being highly defined.

This book provides a general planning masterguide that shows the way to develop a working model of any definable subject matter, like those that have already been developed for the use of planners worldwide: Systematic Layout Planning, Systematic Handling Analysis, Systematic Planning of Manufacturing Cells, Multiple Careers Planning, Systematic Strategic Planning, Results-Based Systematic Operational Improvement, and so on. This objective will be accomplished by introducing the concepts, the process, and the methodology of Planning by Design—the leading-edge methodology on project planning.

The book is written to help graduate students and planners of all kinds understand, put into use, and benefit from a structured but flexible process of planning almost anything with or having a definable subject area. In reality, this book has been designed to be specific, simple to understand, and easy to use.

I owe a great debt of gratitude to all the fellow students who contributed to the development of this book.

Most of all, I am very thankful to Richard Muther and his Planning by Design, potentially the greatest advance of the current century for planners (which is so consistently structured, so well organized, and so logically designed, it is no wonder that project planners are saying: "We never had it so good").

Hakan Bütüner

Author

 **Hakan Bütüner** is a graduate of TED Ankara College. He received his B.Sc. in industrial engineering from Middle East Technical University, his MBA from Bilkent University, and his PhD in engineering management from the University of Missouri–Rolla.

Dr. Bütüner has been active in both his academic and professional lives for several years as a planning and programming manager and as a project manager overseas, and started up and chaired a new venture for bringing different and strong international franchising concepts to Turkey.

Later, he worked as a strategic planning and business development director of Bayındır Holding; as an operations and profit improvement program country manager of Siemens Business Services; and then, as a general coordinator of Bell Holding. During the same period, he also lectured in the Business Schools and/or Industrial Engineering Departments of Bilkent, Bosphorus, Bahcesehir, and Yeditepe Universities.

Dr. Bütüner is currently acting as an affiliate of several U.S. companies in the industrial management and engineering consulting, training, and software solutions field. He is also acting as the founder-president of the Institute of Industrial Engineers—Turkish professional chapter.

During his career path, Dr. Bütüner has participated in several projects both in Turkey and abroad. Additionally, he is a board member of the Institute of High Performance Planners in the United States. He is the author of several publications and books and has been honored by the decision sciences society Alpha Iota Delta.

PART I

PLANNING BY DESIGN (PxD)

Planning is deliberately thinking about the future with the purpose of determining what to do and/or how to get it done.

Researchers have concluded that no comprehensive planning process is currently accepted. Some even maintain that planning is so difficult to formulate that a universal approach may never be realized.

However, systematic planning has been in use since at least the mid-1900s and has been of real benefit in several particular subject areas. And, as you will find in this book, the development beyond systematic planning—Planning by Design, developed by Richard Muther—provides both a comprehensive process and a highly definitive one as well.

1

NATURE OF PLANNING*

Planning is primarily a *mental* activity: we do it with our minds. In beginning to consider planning, there are some principles worth remembering:

- Well-organized planning enhances good ideas.
- The process of planning can be structured.
- An effective planning structure can be applied repeatedly and with confidence and dispatch.
- While the process of planning can be modeled, each plan is situational; each planning situation is different.

Certainly, there are differences in planning situations. Family planning in Sweden is different from population planning in China. Community planning done by a government agency is different from the same planning done by an independent consulting firm. And planning in a familiar or repeatedly encountered situation is less uncertain than under a novel set of circumstances. Still, the process of rational planning is the same, especially for uncomplicated assignments with near horizons.

1.1 Definitions of Planning

Planning has several definitions:

- To determine ahead of time what you should or intend to do
- To think about and decide on a proposed course of action
- To convert desires, goals, objectives, demands, or problems into schemes, decisions, stated intentions, or solutions
- To condition one's mind with future opportunities, obstacles, possibilities, and alternative courses of procedure so the mind

---

* Adapted from Muther (1988).

will be open and quicker to respond to circumstances or to find more creative solutions in subsequent, more definitive planning

As A.A. Milne put it, "Planning is what you do before you do something so when you do it you don't mess it up."

The Institute for High Performance Planners has selected the following: "Planning is the process of determining what we intend to do and how we propose to get it done."

Most decisions are made with some planning—instinctive and immediate, or deliberate and prolonged. In as much as all of us make decisions and think about most of them before deciding, we are all planners, every day, whether we recognize it or not. But some of us are better planners than others. And some are impulsive doers. Sometimes, we even pride ourselves on not doing much, if any, planning.

Before you go somewhere, you should know where you are going. Unless, of course, you want to get lost. Some people are risk-takers who make their activities into adventures or treasure hunts. But most of us find more value in the countering proverb, "Look before you leap." Why? Because planning avoids all the added cost, wasted time, extra effort, frayed nerves, and sometimes irretrievable loss of life that *getting lost* can mean.

As the world becomes more complex, we need to plan better. More is at stake. A high price is paid for things that turn out to be bad, unreliable, or illegal. There is a greater need to integrate people, time, space, and things. There is a greater need to make decisions faster and better.

If you know how to plan, with a method you can apply to most every situation, and if you learn to use it instinctively, you are well on your way to improving your own tight schedule, meeting your company's and family's demands for your time, enhancing your subordinates' and associates' needs for clearer and crisper guidance, and—indirectly—bringing home fatter paychecks and leaner meat for the table.

Well-considered planning is part of the sequence shown in Figure 1.1. Therefore, the logical sequence of getting things done is goals, plans, actions, and results.

These four terms are from a cause-and-effect sequence. Ideally, the results are appropriate to the goals. If not, you may have to establish

The four stages of getting things done

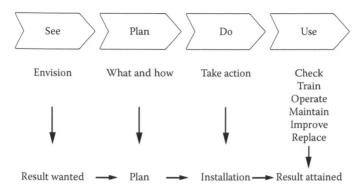

Figure 1.1 Established sequence of getting results. (Reprinted from Muther, R., *Planning by Design*, Institute for High Performance Planners, Kansas City, 2011.)

new goals or shift to another plan and course of action. In this sequence, the goals and results are the ends. The plans and actions are the means to those ends.

1.2 Benefits of Planning

Planning is not only deciding to do the appropriate thing at the appropriate time and space with the appropriate people. Planning helps avoid excessive cost, lost time, wasted effort, and the need to do something over again. In this way, planning is *protective*.

When planning, we are thinking about, analyzing, weighing, appraising, conceiving, and developing ways of doing something. We may create entirely new ways of doing whatever we have set about. Thus, planning is *creative*. It generates potential solutions and fresh opportunities.

Moreover, planning can be a potent propelling force—both for motivation and for enhancing conviction or reassurance. Planning is *motivating*; it promotes enthusiasm for the project and often builds greater loyalty in the individual planners. Additionally, planning is *reassuring*; it promotes confidence in the plan and often builds strong conviction among the planners involved.

When we drive a car, brush our teeth, or clean up spilled coffee, we almost instinctively follow a previously learned procedure. We do it without much thinking. We perform by habit. This book aims to instill a similar responsive type of habit for planning.

The need for a sound guide to planning is clear. If we don't have one, we disadvantage ourselves, because we have to put effort into determining how to plan. Such an approach has significant *economic consequences.* A comparatively small investment in planning time, effort, and money can have a major influence on success, satisfaction, and productiveness during the useful life of the thing being planned.

Additionally, the early phases of planning have the greatest leverage on the outcome. That is when the project is understood and oriented to the situation, and when decisions are made to the extent of planning the time and effort to be invested. For these reasons, learning a systematic way of planning can be a vital skill in business in particular or in life in general.

1.3 Planning Levels

Planning occurs in a series of levels, classified primarily by the size and importance of what you are planning. Figure 1.2 shows several levels from forest to leaf parts.

The significance of this levels-of-planning concept is multiple: each action falls within a larger action in a system-and-subsystem sequence; each plan falls within the larger plan; and each planning assignment itself divides into subassignments-actions within larger actions, plans within larger plans, and assignments within larger assignments.

As a planner, you can seldom do much about a larger, previously made plan. If your job is to plan the clearing of underbrush, you can't go back and change the spacing of the trees. Inevitably, certain conditions surround each planning assignment, conditions you cannot

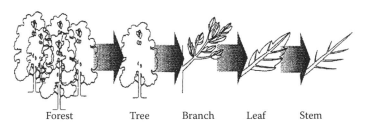

Forest   Tree   Branch   Leaf   Stem

Figure 1.2 Typical sequence of levels of planning. (Reprinted from Muther, R., *High Performance Planning*, Management and Industrial Research Publications, Kansas City, 1988.)

change or control and within which your planning will have to be done.

1.4 Planning Horizons

Plans for the future are made with different levels of detail and for different purposes, depending on how far into the future you are planning. The closer the horizon—and therefore the closer the intended action—the more action-oriented and more specific the planning should be. Conversely, the farther out the planning horizon, the less specific the planning.

As a result, planning for the long term is not concentrated on taking action. Rather, it concentrates on conditioning the mind of the planner: thinking ahead about the opportunities in the remote future and the strategies for the *big picture*. You should not have much confidence in those who admit they don't know where they are going but urge you to hurry up and get there.

On many occasions, however, a plan cannot be devised, because the situation is unclear, or the facts are unknown. You have to make your reconnaissance and attack at the same time. You have to *see what develops*, *play it by ear*, or *take it as it comes*.

And in most emergencies, there is no time to plan. You have to act, not stop to plan. A sudden squall may make staying afloat your immediate objective, and doing things right now becomes your life-saving concern.

On the other hand, too much of anything is not good. Too much time spent in planning can kill a project, for several reasons. Maybe planning eats up all the project time and you get down to *too little, too late*. Maybe planning so annoys impatient doers that they simply want to forget the whole thing. For any of these reasons, too much planning can be very costly.

The choice is not *plan or not plan*, not even too much or too little. Rather, it is: How much planning is appropriate?

2
FORMULA FOR PLANNING*

For discussing the project situations and real results expected, the what-to-do and how-to-do-it courses of action, arriving at several alternatives, and then deciding on the most preferred plan, a proficient way of planning is required.

2.1 Three Aspects

Most everything in life can be classed into one of three aspects or a combination of them. For example, seeing is largely personal or emotional, planning is largely mental, and doing is physical. Figure 2.1 shows the three aspects of most everything.

2.2 Formula for Planning

The three essentials that form the basis of planning as a whole are *discern*, *devise*, and *decide*. This planning triad is important to the planning process, as it forms underlying principles that support the entire effort of planning. A process of discerning, devising, and deciding goes on from the beginning of a planning effort through the actual planning assignment. Figure 2.2 shows the process of planning.

In planning anything, it is necessary to comprehend the information available: the goals or requirements of the project ahead; the context of the project itself; how it is contained in a larger frame of reference; and the particulars: the facts, data, observation, or statements of others that relate to the project. Putting all this together on a grand scale is *discerning*.

Devising is a matter of contemplating or analyzing the influences and information and conceiving or synthesizing a possible approach,

---

* Adapted from Muther (2011).

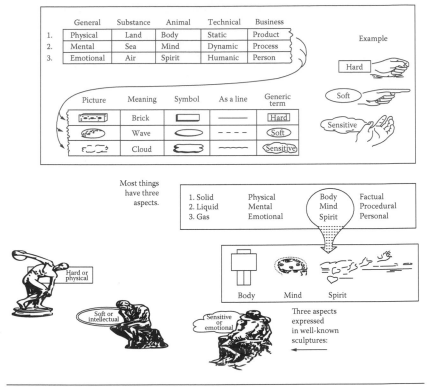

Figure 2.1   Three aspects of most everything. (Reprinted from Muther, R., *Planning by Design*, Institute for High Performance Planners, Kansas City, 2011.)

solution, or plan for the project at hand. It puts focus on the *three fundamentals* of the particular subject area of the project at hand.

Then, the planner moves on to *deciding* the implications of the preferred plan and getting others to accept it. This involves evaluation of the alternative plans devised, with the decision typically resting on *hard*, *soft*, *and sensitive aspects*.

These three essentials are repeated in virtually every phase of planning. In most cases, they are almost unconscious—an experienced planner goes through them in the planning process without having to think.

In broad terms, "planning" is defined as the process of determining what we (I, you, they) intend to do and/or how we propose to get it done.

Further, we need to realize that there is a formula for planning. The formula may be expressed as follows:

$$P = 3D = D_1 + D_2 + D_3$$

Where P = the planning process
D_1 = discern = discover and understand
D_2 = devise = diagnose and develop
D_3 = decide = select and accept

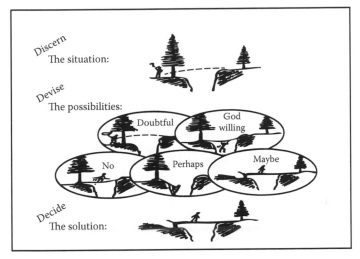

For some, this is a new idea: that planning boils down into three basic activities. That too is reality.

Figure 2.2 The process of planning. (Reprinted from Muther, R., *Planning by Design*, Institute for High Performance Planners, Kansas City, 2011.)

3

SYSTEMATIC PLANNING*

As defined earlier, planning is the process of determining what we intend to do and how we propose to get it done. *Systematic planning*, then, is doing this planning in an orderly, rational, organized, and methodical systematic way.

3.1 Systematic Planning

By its nature, planning involves a sequence of activities. And the sequence fits into the larger sequence of getting things done, as shown in Figure 3.1:

- Understand the project and how we plan to do the determining: SEE (*Plan the planning*).
- Determine what we intend to do: PLAN (*Do the planning*).
- Determine how we propose to get it done: PLAN (*Plan the doing*).
- Take action: DO (*Do the doing*).

This pattern is still not a generally accepted pattern for planning, as systematic planning means different things to different people. This is one reason why a different approach, known as *Planning by Design*, is introduced in the following chapter.

3.2 Comparison with Scientific Management

Scientific management is defined as the planning, organizing, leading, and controlling of work of any kind. It sets standards, measures or establishes norms, and/or sets best-practice procedures, and then,

---

* Adapted from Muther (2011).

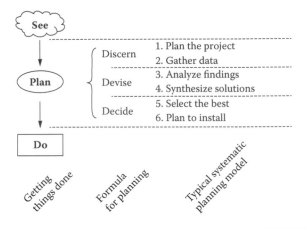

Figure 3.1 Systematic planning. (Reprinted from Muther, R., *Planning by Design*, Institute for High Performance Planners, Kansas City, 2011.)

it measures the performance and takes corrective action when that performance is outside the acceptable limits of the standards. With this definition, much of modern management is *scientific*.

On the other hand, most of the planning is not scientific, for planning is always about the future, and much more of the future is unknowable. Still the process of planning can be highly *systematic*.

Systematic planning's development is ongoing. It is very much like the movement of scientific management during the last century, which took a long time to be understood and accepted. Today, those concepts of tooling-up to produce procedures to guide recurring activities are being extended into the largely mental process of planning. With systematic planning, we are able to provide working models for the planning of projects well before we know what the specific project even is. Comparison of systematic planning with scientific management is shown in Figure 3.2.

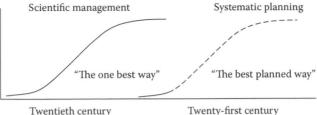

Scientific management
- The objective measurement of performance
- Improving the doing of work
- The one best way

Systematic planning
- The orderly determination of intentions
- Improving the planning of work
- The best planned way

Figure 3.2 Relationship of systematic planning to scientific management. (Reprinted from Muther, R., *Planning by Design*, Institute for High Performance Planners, Kansas City, 2011.)

3.3 Degrees of Planning Refinement (Table 3.1)

Table 3.1 Understanding the Refinements of Planning

PLANNING	SYSTEMATIC PLANNING	PLANNING BY DESIGN	HIGH-DEFINITION PLANNING BY DESIGN
"The largely mental process of determining what we (you, I, they) intend to do and/or how we propose to get it done."	Planning (as defined) with an orderly way of determining.	PxD for short—is systematic planning with a pre-designed process of determining for the particular subject area of concern.	Hi-Def PxD for short—is where each step of the process has a pre-designed form of output, and a pre-designed key document leading to that output.
Note: • Largely mental • Process • What and/or how • Intended doing	Note: • Orderly way (or structured, or patterned way) • Determining	Note: • Pre-designed process (or working model) • Particular • Subject area	Note: • Each step • Key document • Form of output
Usage: Everyone every day.	Usage: Most planners follow one of the many processes of planning considered to be orderly.	Usage: Many people use/apply particular subject-area working models, but few develop/derive the particular-subject working model they use.	Usage: Few people have had the time and interest to develop working models for the particular subject area(s) of their interest.

(Continued)

Table 3.1 (Continued) Understanding the Refinements of Planning

PLANNING	SYSTEMATIC PLANNING	PLANNING BY DESIGN	HIGH-DEFINITION PLANNING BY DESIGN
Its problem: We have to think.	Its problem: We have to think in an orderly way.	Its problem: We have to pre-design the orderly way.	Its problem: We have to pre-determine the form of output, and key document for each step of the orderly way.
One line: "Most profitable activity of mankind."	One line: "The major advance of the 21st Century."	One line: "The gold standard of planning processes."	One line: "The swiss army knife of systematic planning."
Comment: Of course, everyone makes plans—every day—without much concern about how they do it. Its base requirements is that you have to think.	Comment: But there is no comprehensive "orderly way" that is generally accepted.	Comment: Several subject areas have working models that are very popular—SLP, SHA, et al. But few of us know how to design/ develop the PxD working model.	Comment: High-definition planning by design is most helpful when developing the working model.

Source: Muther, R., *Planning by Design*, Institute for High Performance Planners, Kansas City, 2011.

4

ESSENCE OF PLANNING BY DESIGN[*]

Planning is primarily a mental activity; we do it with our minds. The mind has a generalized way of thinking, diagnosing, planning ... and if we work with it, we get a generalized masterguide for planning any project. Figure 4.1 shows the similarity of mental activity to systematic planning.

This model is for rational planning. It is directly applicable to the vast majority of all planning. Additionally, it serves as a point of departure and basic guide for any planner who doesn't quite know how to proceed.

Dick Muther has spent more than six decades and 2000 projects distilling his systematic planning processes, named *Planning by Design* (PxD) methodology. Planning by Design has more than one meaning:

- Loosely, it means planning *with deliberate intent.*
- Systematically, it means planning *with an orderly structured process.*
- And particularly, it means planning *with a pre-designed working model.*

PxD is determining what you intend to do and/or how you propose to do it by following a pre-designed process that tools up the *way* of determining before you know what your specific project is.

In short, PxD is a methodology that *makes nonroutine projects routine.*

The essence of PxD is that the planner applies a working model for the subject area of the project at hand (such as designing a

---

* Adapted from Muther (1988) and Muther (2011).

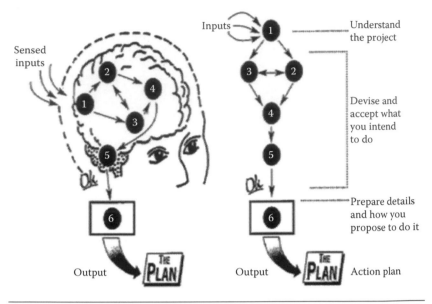

Figure 4.1 Similarity of mental activity with systematic planning. (Reprinted from Muther, R., *Planning by Design*, Institute for High Performance Planners, Kansas City, 2011.)

manufacturing cell, creating an office layout, planning a family vacation, planning a career, etc.) The working model is

- Consistent with the *formula for planning*
- Particular for the *fundamentals of the subject area*
- Appropriate to the *size* of the project

4.1 Subject Area and Its Fundamentals

The *subject* is the topic or theme of the project at hand. This is not the name of the project, or a general description of the subject area, or a definition of planning or the planning process, or the purpose or the result wanted. *It is the name of the subject area with which the project is dealing.*

For example, if your project is to plan a quality assurance program, it is easy to interpret this as improving the quality of the products. But the subject area is to set up a quality assurance program, and the project is determining what to do to plan such a program and/or how to get it established.

In PxD, defining the subject area of the project at hand is vital. Now, you are forced to work with a particular subject area and a particular set of fundamentals to support that subject area.

Fundamentals are the rudimentary elements that are important to, always involved in, and singular, if not unique, to the particular subject area.

There are *three key fundamentals* special to each subject or discipline. Once you have found or developed these, your planning has a structure to support the many different ideas you may try out in the course of creating the best plan.

In understanding cosmic concepts, mankind has often turned to triads. From early times in the world's religions, triads of divinities have been important, perhaps, some anthropologists believe, because three is the number of a primal family: man, woman, and child. The belief systems of Buddhism, Hinduism, Taoism, and Christianity all have important places for a trinity.

In seeking the key fundamentals of each particular subject area, ask the question: "What are the basic things that are always involved in making workable plans in this subject area?" Omit things that may be present but that are not the principally defining factors.

What are the basic things always involved—those we are always dealing with—in, for instance, a materials handling or transport plan? They are

- *Materials* to be moved (goods, items, tangible things)
- *Moves* to be made (from pick-up to set-down, usually including both)
- *Method(s)* of moving (handling/transport equipment, container and move routes)

Finding the key fundamentals in a particular area, topic, or discipline can be one of the most interesting challenges of planning. Discovering the key fundamentals and building them into your *devising* will speed planning, make it clearer, and provide greater conviction for you and more confidence for others. Figure 4.2 shows the fundamentals and the formula together.

4.2 Size of Project

There are multiple sizes of planning projects, and the degree of complexity affects the *seeming size* of the project. So, it is customary to speak of the mix of planning projects as

Focus on the fundamentals of the subject area at hand, and tie them to the formula

<u>Discern</u> A and B   1. Define the subject area

2. Determine its fundamentals

Relate A and B   3. Relate fundamentals A and B
and <u>Devise</u> Cs
4. Devise alternative Cs

<u>Decide</u> which   5. From flipping a coin to using a
alternative plan   weighted-factor comparison
is best

The fundamentals (A, B and C) relate to the formula as follows:

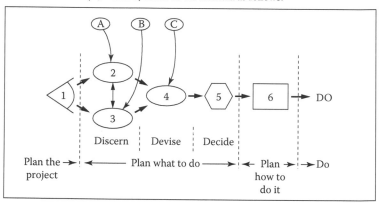

Figure 4.2   Fit the fundamentals and the formula together. (Reprinted from Muther, R., *Planning by Design*, Institute for High Performance Planners, Kansas City, 2011.)

- *Small projects:* or subprojects of a larger one.
- *Medium-size projects:* need to be planned in phases, applied in a larger pattern, and approved at the completion of each phase.
- *Large or complex projects:* physically large; extra long time horizon; decisions to be made fall into two or more subject areas; subject areas are interrelated; implementation may take place

in stages over a long period of time or is itself complex and will need its own set of stages.

4.3 Structured Pattern of Three Masterguides

Dick Muther has spent more than six decades and 2000 projects distilling his systematic planning processes into three master-guides. PxD has three masterguides—short form, full, and extended—one for each general size of project. Figure 4.3 shows the three size structures. Each masterguide provides a structure for a working model:

- *Short-form PxD:* Single-phase six-step masterguide for short-form working model. Useful for *small* projects or *Phase III of larger ones.*
- *Full version PxD:* Four-phase planning project with repeating masterguide in five sections of Phases II and III. Useful for *medium-size* projects.
- *Extended version PxD:* Six-plus phases of major project with lead component, parallel subject areas, and repeating patterns in Phases II and III. Useful for *large* or *complex* projects.

4.3.1 Full Version PxD

Four-phase planning project with repeating masterguide in five sections of Phases II and III. Useful for medium-size projects, such as layout planning of a factory or strategic planning of a corporation.

4.3.1.1 Four Phases of the Planning Pattern Any planning assign-ment or project can typically be divided into four phases, as shown in Table 4.1.

The middle two phases are the heart of the planning process. They provide a sequence of system and subsystem planning. From the whole to the parts; from the big to the small; from the general to the specific—this top-down sequence that progresses through Phases II and III of planning is important for several reasons:

- You don't spend time and money on the details till you know the whole is acceptable.

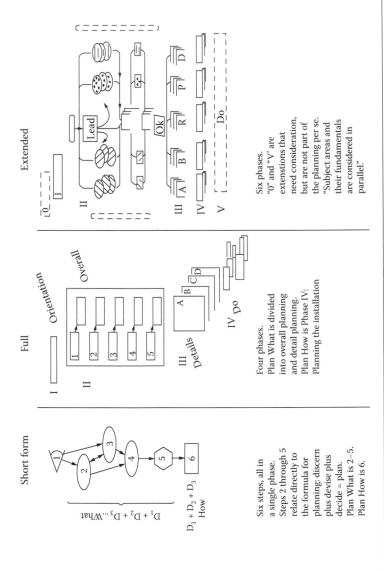

Figure 4.3 Three appropriate-size structures. (Reprinted from Muther, R., *Planning by Design*, Institute for High Performance Planners, Kansas City, 2011.)

Table 4.1   Four Phases of the Planning Pattern

NUMBER	NAME	BASIC ACTION	OUTPUT
I	Orientation	See it whole	The project understood in its surroundings
II	Overall	Plan it sound	The solution in principle or long-term solution
III	Detail	Make it real	The solution(s) in detail or short-term plans
IV	Do/Act	Follow it through	The plan accomplished/implemented

- You don't work on the fragments and expect them to fit together into an integrated plan.
- You don't charge off into specifics and later find they are not compatible with the more important whole.
- You don't limit your big-picture imagination with constraining details in the beginning. The details of your plan come after you have envisioned the project as a whole.

The phases of planning come in *sequence*. Figure 4.4 shows the framework of planning phases. In addition to coming in sequence,

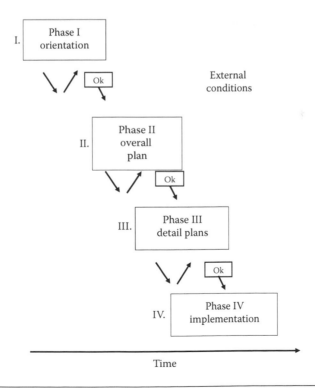

Figure 4.4   Framework of planning phases. (Reprinted from Muther, R., *High Performance Planning*, Management and Industrial Research Publications, Kansas City, 1988.)

for best results, the phases should *overlap* in time. You are think-ing ahead in your planning as you work in each phase. You are already planning certain critical details before you accept the overall plan.

Additionally, for effective planning, there should be an okay, accep-tance, or *approval* at the end of each phase. This allows you to move ahead with reasonable assurance. It permits others who may have dif-ferent viewpoints or greater wisdom or deeper investment to share their views. If you get new input after acceptance, you can go back and reopen a prior phase, because you are still *only on paper* till Phase IV is underway.

The conclusion of each phase acts as a basic check point. So, they make logical *division points* in the planning and scheduling of the planning work itself. Here, the planner and the management verify the status of the assignment, adjust its conditions, reject or modify the planning done so far, assess the time and costs and likelihood of success, and generally agree to proceed seriously with the next phase, to hold off till later, or to scrap the assignment.

4.3.1.1.1 Orientation   There is always a larger, higher-level, or lon-ger-term plan within which your current project or planning assign-ment must fit. These larger, external, *environmental* conditions become the assumptions, boundaries, or external considerations within which your project will have to develop.

- Understand the *objective(s)* of the assignment—its purpose, goals, priorities, or requirements.
- Determine the *external conditions*—any larger prior plans; overall strategies; basic assumptions or other surrounding legal, financial, technological, political, organizational, or economic constraints to be reckoned with.
- Clarify the planning *situation*—who is responsible for doing what by when; where to work; others involved; any mandatory aspects of the assignment or any constraints under which the planning is to be done.
- Determine the *scope*, the *extents*, or *boundaries* of the planning—how much, how deep, and the expected form of the anticipated output (or deliverables).

Table 4.2   Going from Phase II to Phase III Planning

PHASE II	PHASE III
The forest	Trees, fire lanes or roads, undergrowth, sawmills …
Automobile	Plans for chassis, body, engine, transmission, wheels …
Design of watch	Design of case, hands, face, movement …
City plan	Throughways, park, commercial zones, residential areas …

- Make a *plan* for the planning—a schedule of tasks or activities to be undertaken.

4.3.1.1.2 Overall plan   The aim in Phase II is to determine a general, total, or long-term plan such that *in principle*, it is basically sound, meets the objectives of the project, and integrates with the external conditions.

Phase II is the actual creation of the plan. Here you will get and *understand* the inputs, *devise* and develop the concepts, and select and *accept* the solution or plan. The output of Phase II is an *overall plan*. This is a plan for the whole situation or total system, as identified in Phase I.

Phase II involves the planning pattern in its five sections. The overall plan is larger, longer range, or more comprehensive then the following Phase III planning. The move from overall planning to detail planning is typified by expressions such as

- Whole to parts
- System to subsystems
- General to specifics
- Long term to short term
- Principle to practice
- Policy to procedures
- Plan to design

4.3.1.1.3 Detail plans   Phase III repeats the same essential planning process, but it does so at a more specific level, as shown in Table 4.2.

Note that the output of Phase II is for *detail plans*—plural. There is usually more than one. Typically, these are for smaller areas, shorter times, or more definitive elements of the whole, or for more particular skills and disciplines. Phase II planning, for example, might be a marketing plan for the ABC Company. Within that, Phase III planning would involve a plan for direct marketing, a plan for trade show marketing, and a plan for space advertising. All of these would meet the objectives of the company, which would have been taken into account in Phase I. Characteristics that distinguish Phase III from Phase II include all or many of the following:

- More details, more specifics
- Several plans (compared with only one in Phase II)
- Smaller spaces or areas considered
- Shorter horizons
- Data or input better understood
- Different persons doing the planning
- Planners' skills more particular, less comprehensive
- More man-hours required for planning

Phase III also supports the plan or solution of Phase II and in a real sense assures that it will work. Phase III explores and examines the *parts* of Phase II. Unless there is some rejection of the detail planning, these plans should work.

4.3.1.1.4 Implementation   Phase IV is probably the most rewarding part of planning. Phase IV is making the plans happen. This phase is dedicated to carrying out the plan.

- Verify the approval(s), funding, and responsibilities to carry out the plans.
- Prepare for implementation—who does what, when, and how.
- Do, act, direct the physical implementation; take the necessary action to get the plans executed.
- Control time, cost, and correctness of the implementation, correcting as necessary and/or appropriate; then release to operating persons.
- Follow up, check out, debug to be sure the actions taken do indeed meet the plans and accomplish the objective(s).

Whether or not they are directly involved, planners will share the credit or blame for the way the plans are carried out.

It is very common for the planners to turn the project over to the implementers. Architects and engineers, for instance, plan the building; a contractor gets it built. In such a division of effort, the planner should be responsible for seeing that plans are carried out.

4.3.1.1.5 Repetition of the pattern　Phases II and III are the heart of the planning. Phase II establishes the overall plan or the solution in principle. Phase III establishes plans or solutions in detail. But the sequence of planning is essentially the same in both phases. The pattern is the same in Phases II and III. However, its application is to different levels of detail and often to different time horizons. But the planning pattern repeats, as is shown in Figure 4.5.

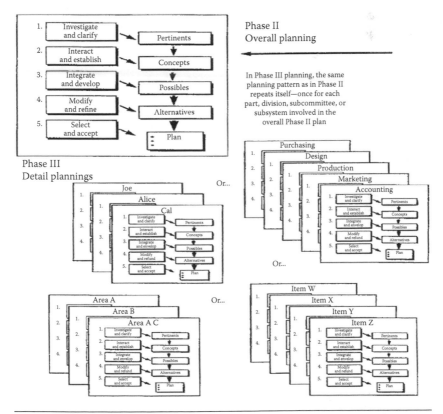

Figure 4.5　Repetition of the planning pattern. (Reprinted from Muther, R., *High Performance Planning*, Management and Industrial Research Publications, Kansas City, 1988.)

This example shows how Phase II is divided (in Phase III) into areas or products or committee chairpersons or operating departments.

4.3.1.2 Five Sections of the Planning Pattern   In Phases II and III, there is a logical order. They are divided into five sections, as is shown in Figure 4.6. In rationale, the planner typically progresses in this order:

1. INVESTIGATE the context, inputs relative to the key fundamentals, and influences (parameters, existing and future conditions within the particular project/assignment and its planning), ORGANIZE the project, and make it understood.

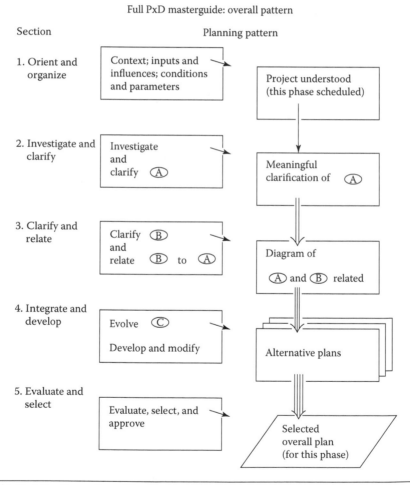

Figure 4.6   Five sections of the planning pattern. (Reprinted from Muther, R., *Planning by Design*, Institute for High Performance Planners, Kansas City, 2011.)

2. INVESTIGATE and CLARIFY the first key fundamental with the pertinent features.
3. CLARIFY the second key fundamental and RELATE the first two fundamentals to each other through analysis and synthesis and/or inductive reasoning.
4. INTEGRATE with these base concepts the third fundamental and such considerations as people, time, place, and cost, and DEVELOP into possible plans.
5. EVALUATE the developed alternative plans, SELECT from the alternative plans, and APPROVE the selected one(s).

4.3.2 Short-Form PxD Pattern

Short-form planning is often called *problem-solving*. The short form condenses the four phases into six steps, as is shown in Figure 4.7, and combines Phase II and Phase III. It is applicable to short, smaller planning assignments or situations, and in these cases, it is used instead of the framework of phases and the full planning pattern.

In simple planning assignments—the ones that most of us encounter most often—such as planning of a manufacturing cell, planning of a career, and so on, the short form will suffice.

1. Orient and Organize
- Understand the project: What? Why? Who? When? Where?
- Understand the purpose(s) and objective(s), external conditions, situation(s), scope, budget, and form of output.
- Understand and document planning and people issues.
- Identify the project's subject area and identify its three key Fundamentals, A, B, and C.
- Make a schedule for planning the project.

Output: The project understood and its planning sceduled.

2. Investigate and clarify fundamental A
- Clarify the meaning of A further.
- Determine what data is needed, how to get it, from what sources, and with whom.
- Think of inputs as (a) physical/tangible, (b) procedural/mental, and (c) personal/emotional.
- Go after data—and organize, clarify, sort, sequence, compare, and otherwise analyze and/or visualize the data.

Output: Meaningful clarification of A data.

3. Clarify fundamental B and relate to fundamental A
- Follow much the same procedure as in Step 2. Clarify further the meaning of B.
- Determine what data you will need, how you propose to get it and from what sources and with whom.
- Think of inputs as (a) physical/tangible, (b) procedural/mental, and (c) personal/emotional.
- Go after data and organize, clarify, sort, sequence, compare, and otherwise analyze and/or visualize the data.
- Then relate the first two fundamentals, A and B.

Output: A tangible visualization of the related first two fundamentals.

4. Integrate fundamental C and develop alterntives
- Clarify further the meaning of C. Gather any additional data.
- Join C with A and B in ways that will meet the project obejctive(s).
- Resolve any as-yet-unresolved issues identified in Step I or later.
- Develop the concepts into at least two viable alternative what-to-do plans or courses of action.
- Explain how each alternative will work.

Output: Two or more viable alternative what-to-do plans.

5. Select and accept the best
- Identify each alternative as defined in Step 4.
- Obtain from appropriate in-house and outside sources the costs of services and/or equipment necessary.
- Determine the operating costs for each alternative plan and prepare a worksheet showing the comparable costs for each alternative.
- Also, on a separate copy of a similar worksheet, make a comparison of the intangible benefits and risk of each alternative.
- Compare the alternatives: select the best: get others to accept it.

Output: An accepted decision of what to do.

6. Detail and prepare to do
- Check and consider any details or specifications that are needed to support your selected what-to-do plan.
- Make a list of things you and/or your group must do to accomplish the plan's implementation or installation.
- Determine any special resources required.
- Assign responsibility for each task to you and/or a member of your group.
- Schedule each task.
- If the planner is also the doer's/he will want to post periodic performance data against the scheduled tasks and take appropriate action.

Output: An implementation plan of how to get the what-to-do of Step 5 completed, approved, and ready to DO.

Reprinted from Muther, R., *Planning by Design*, Institute for High Performance Planners, Kansas City, 2011.

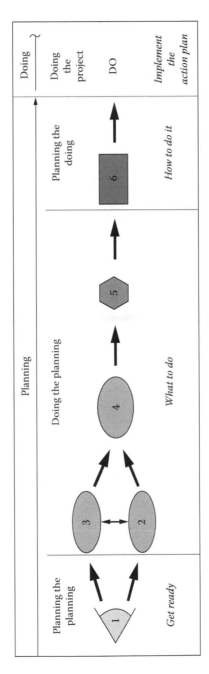

Figure 4.7   Short-form planning pattern. (Reprinted from Muther, R., *Planning by Design*, Institute for High Performance Planners, Kansas City, 2011.)

4.3.3 Extended Version PxD

For planning large and complex projects, PxD has an extended version, shown in Figure 4.8, with

- More phases
- Explicit recognition of existing plans or conditions and their long-term implementation programs

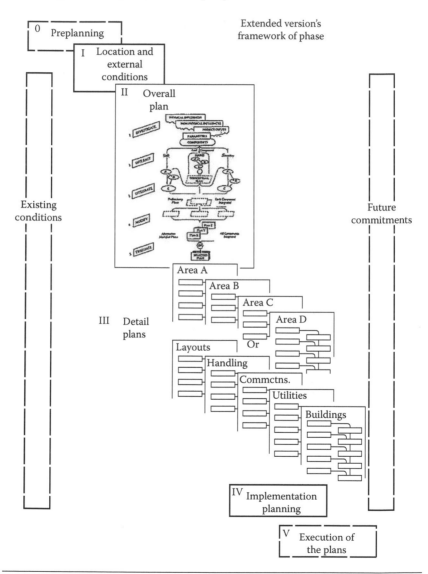

Figure 4.8 Extended version PxD. (Reprinted from Muther, R., *Planning by Design*, Institute for High Performance Planners, Kansas City, 2011.)

- The ability to plan two or more subject areas or components in parallel
- A way to establish the lead or primary component that will drive the planning

The full PxD masterguide does not suffice, since it is basically designed to address a single subject area. So, full PxD is stretched to accommodate multiple components of subject area, large physical size, prior decisions, separate implementation planning, and so on.

4.4 Specific, Particular, and General

PxD provides a *general* planning *masterguide* that identifies a way to *plan the planning* for *any definable subject area* as a *project*, from designing a manufacturing cell, to creating an office layout, to planning a family vacation, to planning a career, and so on.

A working model, shown in Figure 4.9, is therefore a procedure in words, or a pattern in diagram, or an intellectual map that follows the structure of a generalized model but is expressed in terms, techniques, and methods of the project at hand.

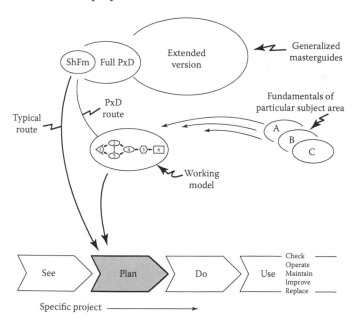

Figure 4.9 Specific, particular, and general. (Reprinted from Muther, R., *Planning by Design*, Institute for High Performance Planners, Kansas City, 2011.)

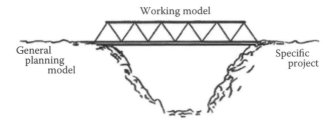

Working model

General
planning
model

Specific
project

Bridging the gap between a generalized planning model and a specific project,
with a working model, is comparable to ...

the military sequence:

Strategic plan   →   Tactical plan   →   Action plan

or the business sequence:

Policy   →   Procedure   →   Practice

Figure 4.10   Working model bridges the gap between generalized planning and specific project.
(Reprinted from Muther, R., *Planning by Design*, Institute for High Performance Planners, Kansas
City, 2011.)

By developing a PxD working model for a particular subject area
of primary interest, we have the benefit of *an already half-made plan*
for any specific project in that subject area. Therefore, PxD bridges
the gap between the generalized master guide and the specific project
with a working model, as shown in Figure 4.10.

As an example, Figure 4.11 shows a specific six-step pattern by
which a small project in the *particular* subject area can follow a *working model* for that subject area.

4.5 High-Definition PxD

When each step or a section of a working model is supported by the
particular form of *output* and *key document* leading to it, we have what
is called a High-Definition Working Model.

For instance, the form of output for Step 3 of competency-
based training needs analysis (Chapter 6) is described as a *position
and competency matrix*, shown in Table 6.5. Further, the key documents of Step 3—the document that supports or leads to the form of
output—are shown in Tables 6.3 and 6.4.

Understand working model

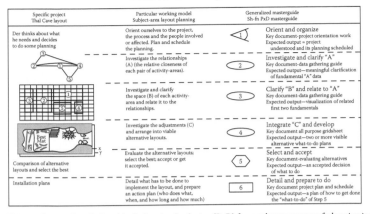

Specific project Thal Cave layout	Particular working model Subject-area layout planning	Generalized masterguide Sh-fn PxD masterguide
Der thinks about what he needs and decides to do some planning	Orient ourselves to the project, the process and the people involved or affected. Plan and schedule the planning.	**Orient and organize** Key document-project orientation work Expected output = project understood and its planning scheduled
	Investigate the relationships (A) (the relative closeness of each pair of activity-areas).	**Investigate and clarify "A"** Key document-data gathering guide Expected output—meaningful clarification of fundamental "A" data
	Investigate and clarify the space (B) of each activity-area and relate it to the relationships.	**Clarify "B" and relate to "A"** Key document-data gathering guide Expected output—visualization of related first two fundamentals
	Investigate the adjustments (C) and arrange into viable alternative layouts.	**Integrate "C" and develop** Key document all purpose gridsheet Expected output—two or more visible alternative what-to-do plans
Comparison of alternative layouts and select the best	Evaluate the alternative layouts; select the best; accept or get it accepted.	**Select and accept** Key document-evaluating alternatives Expected output—an accepted decision of what to do
Installation plans	Detail what has to be done to implement the layout, and prepare an action plan (who does what, when, and how long and how much)	**Detail and prepare to do** Key document project plan and schedule Expected output—a plan of how to get done the "what-to-do" of Step 5

Primary feature that distinguished planning by design (P×D) from other processes of planning is that it pre-designs a WORKING MODEL. See the center column above.

The P×D working model is:

1. Compatible with the formula for planning

$$D_1 + D_2 + D_3 \longrightarrow \text{Plan}$$

2. Particular to the fundamentals of the subject area

$$\text{A related to B} \longrightarrow \text{C}$$

3. Appropriate to the size of the project at hand

Short form, full, extended

Figure 4.11   Example specific six-step pattern. (Reprinted from Muther, R., *Planning by Design*, Institute for High Performance Planners, Kansas City, 2011.)

PART II
SAMPLE
SYSTEMATIC
METHODOLOGIES

Rather than explaining Planning by Design (PxD) in detail only, this second part shows how PxD is applied for developing working models in any type of subject area.

This part demonstrates the application of creative systematic planning in real-life cases and developing practical working models on different subject areas together with my dear students (i.e., even by professionals not skilled in PxD). Some of the working models are structured on the short form and some on the full version of the PxD pattern; however, most of them are structured without being highly defined.

5

SYSTEMATIC MARKETING INNOVATION PLANNING (SMIP)

HAKAN BÜTÜNER AND ONUR DEMİREL

This chapter outlines a systematic methodology for planning a marketing innovation through a step-by-step procedure. Marketing innovation planning is an applicable methodology for figuring out the different aspects of the interested market and accordingly finding out the most feasible and efficient innovative marketing mix solutions while considering the evolution of the relevant market and its customers' expectations.

5.1 Introduction to SMIP

While innovation is accepted as today's most desirable marketing strategy, many businesses still refuse to focus on innovative research projects. The main reasons are the uncertainty of the innovation process and the unpredictable marketing results of the innovated solutions.

Innovation is the process of making changes to something established by introducing something new that adds value to customers and contributes to the knowledge store of the organization (O'Sullivan and Dooley 2009).

The main criteria for considering a new idea as an innovative one lie in its marketability. The new idea should be feasible and applicable. The crucial part of the innovation is that it should get the attention of the targeted customers and persuade them to pay for the new idea. Then, this new idea can be termed *innovative*.

Taking this discussion into consideration, we can say that innovation can be done by differentiating the marketing mix, which includes product/service design, branding, management of sales channels, customer processing, packaging design, management of partnerships, and pricing within the business model. Accordingly, to come up with effective innovations, we should understand the geographical, demographical, behavioral, and psychological characteristics, sizes, and growth rates of the market segments.

Here, we provide a working model for innovation planners to take the necessary actions for planning their marketing innovations. The SMIP is illustrated in Figure 5.1.

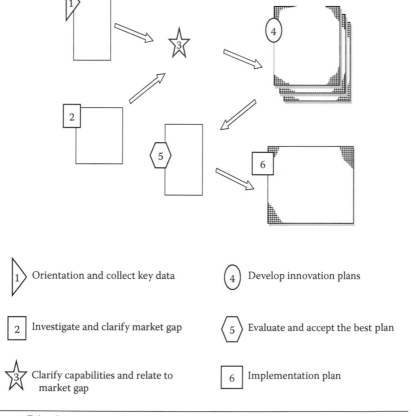

Figure 5.1   Systematic marketing innovation planning.

5.2 Three Fundamentals

Three fundamentals of marketing innovation planning are market gap, capabilities, and innovations.

5.2.1 Market Gap

Today, instead of treating people simply as consumers, marketers need to consider them as human beings with minds, hearts, and spirits. Increasingly, consumers are looking for solutions to their anxieties to make the world a better living place. In a world of confusion, consumers look for companies that address their social, economic, and environmental expectations in their mission and values. They look for both functional and emotional fulfillments in the products and services that they choose (Kotler 2010).

Accordingly, marketers should know that targeting a mass market is always more costly, less effective, and much more risky. They should also know that customers should be segmented and served in various ways based on their geographical, demographical, psychographic, and behavioral characteristics. Dividing markets into segments helps to discover the unsatisfied consumer needs by which incremental innovations can be generated more easily. Therefore, segmentation can help marketers to develop more sustainable customer-driven marketing strategies (Jaman 2012).

Marketers try numerous techniques to identify the most profitable market gap within the targeted segments. Consumer behaviors and market conditions have always been observed from different perspectives through surveys, interviews, experiments, observations, statistical analysis, and so on.

Marketers should also ask their prospects the right questions and listen to them carefully. They should focus on the problems that their prospects are experiencing, not only the features of their products or services (Renvoise 2007).

5.2.2 Capabilities

Every company has different capabilities. They can be briefly categorized as financial, managerial, human resources, sales and marketing abilities, software and hardware capabilities, research and development capabilities, know-how, networking, and corporate culture. All of them together affect the value and quality, development capability, and customer satisfaction of the product/service.

The company's core capabilities, and those that it can develop or acquire, limit what it can accomplish. However, a broader view brings in the notion of distinctive capabilities. These broader capabilities include an organization's "architecture," and this embraces the network of relationships within, or around, the company. These relationships may cover customers, suppliers, distributors, or other companies engaged in related activities (Trott 2005).

Therefore, holding and managing all functions by themselves is not a "must" for today's smart companies. Companies with this new outlook are ready to assign even strategic functions to third parties.

Accordingly, the capabilities of the company and its network of relationships should be analyzed to understand whether feasible solutions exist prior to assigning substantial resources to filling the determined market gap.

5.2.3 Innovations

Once the future customer needs, demands, purchasing behaviors, and company capabilities are clarified, then we are ready to provide innovative solutions by considering their marketabilities.

Therefore, innovation is the application of new or better solutions that create new markets or generate efficiency in an existing market.

"An innovation, to be effective, has to be simple and it has to be focused. It should do only one thing, otherwise, it confuses. If it is not simple, it won't work" (Drucker 2002).

As the market space gets crowded, prospects for profits and growth are reduced. Products become commodities, and cutthroat competition turns the red ocean bloody. Blue oceans, in contrast, are defined by untapped market space, demand creation, and the opportunity for highly profitable growth (Kim and Mauborgne 2005).

By the creation of innovations, companies may desire to get themselves out of price competition, increase their market share, or enlarge the total market size, but there is always a risk for innovative projects because of the obscurity of being new to the market. The failures of innovative projects are mostly grounded in inadequate marketing research, overestimated market sizes, budget exceeding development costs, wrong pricing, poorly designed products or services, incorrect market positioning, incorrect advertising, wrong decisions on sales channels management, ignorance of customer perspective, or early timing of innovation.

To be successful, innovative products must take into account opportunities provided by new technology and materials, on the one hand, but they must not lose sight of the customer, on the other. If designers lose touch with the values, beliefs, and needs of the marketplace, it may be difficult to get new products accepted, and if companies do not have an integrated project team, problems are likely to occur even before product launch (Jerrard 2004).

Applying the traditional customer research methods in an innovation-generation process is surely useful. But as these kinds of research methods are grounded only in customers with limited visions or knowledge, the results may not go beyond incremental innovations, also called *demand-pull innovations*. Most customers do not have the kind of imagination that would lead to the development of such major products as the computer and the automobile. But there could be exceptions; some talented customers might foresee the future. So, it is important to identify how customers respond to products, new or old. Perhaps, their responses could lead to a radical innovation or a breakthrough, or more likely, to incremental innovations that could make a difference to the company's current market placement (Samli 2011).

To analyze the role of customer-related proactiveness in the process of developing radical innovations demands a deeper understanding of the phenomenon. This naturally guides the methodological choices for the empirical research (Sandberg 2008).

5.3 Six Steps

In the framework of this planning pattern, you need to pass through the following six steps to develop the best marketing innovation plans.

5.3.1 *Collect Market Data and Orientation*

To create values for customers and build meaningful relationships with them, marketers should first gain deep insights into their needs.

Companies use such customer insights for developing their competitive advantage.

The objective of market data collection is to identify the opportunities while executing the innovation process. Marketers should understand the consumers' motives, purchasing behavior, and competition levels and the evolution of the market before they start to innovate.

A company is called *market oriented* if it systematically monitors the development of the market (both customers and competitors) and changes its products and services in such a way that the requirements of these developments will be met (Trail 1997).

Dividing markets into segments and defining each segment in terms of market size, growth rate, and profitability enables marketers to find the right market segments to invest in. If there are market segments that are not large enough, or are not measurable, or even consist of consumers who are not reachable, the company must avoid investing in them.

5.3.1.1 Market Segmentation Market segmentation is the starting point of marketing research. Companies have to define whom to serve. Dividing markets into customer segments and selecting the target segment reduces the complexity and increases the effectiveness of the marketing efforts. This decision also makes it possible to create more valuable propositions and to generate the highest rates of customer satisfaction and profitability.

A market should be segmented according to its geographical, demographical, behavioral, and psychographic characteristics (Kotler and Keller 2012).

- *Geographical market segmentation:* This categorizes customers according to their geographical characteristics, such as nations, states, regions, cities, or neighborhoods. A company can target one or more geographical areas while taking other market segmentation characteristics into consideration.
- *Demographical market segmentation:* This categorizes customers according to their demographical characteristics, such as age, gender, occupation, socioeconomic status, marital status, education, ethnicity, religion, nationality, family size, and so on.

- *Behavioral market segmentation:* This is based on the customer's attitudes within the usage characteristics of a product or a service. Marketers believe that behavioral variables, such as benefits, user status, rate of usage, usage size, price sensitivity, buyer-readiness, and loyalty, are important for the categorization.
- *Psychographic market segmentation:* This is based on two customer principles: personality and lifestyle. Psychographic profiling is often used as a supplement to geographical and demographical segmentation when they do not provide the marketer with sufficient understanding of customer behavior.

5.3.1.2 Investigation of Competition   According to the basics of marketing, any company should provide its customers with better values and feelings of satisfaction than its competitors. In this case, within the targeted segment, the company must investigate its competitors in various aspects.

The marketers of the company should figure out the competitive advantages and weaknesses of each challenging competitor (see Table 5.1).

5.3.1.2.1 Product-/Service-Based Competition   A product/service aims to satisfy the need of a targeted segment. Each company competes with a product/service range that consists of a combination of relevant product categories (Trott 2005), while each category is made up of similar products with different specifications (see Table 5.2).

5.3.1.2.2 Branding-Based Competition   A brand is a name, a sign, a symbol, a design, or a combination of all these that identifies the maker or seller of a product or service. Customers view a brand as an important part of a product, and branding adds value to a product.

Table 5.1   Investigation of Competition

	STRENGTHS	WEAKNESSES	COMPETITIVE ADVANTAGES
Competitor 1			
Competitor 2			
Competitor 3			
Competitor 4			

Table 5.2   Product-/Service-Based Competition

		PHYSICAL	FUNCTIONAL BENEFITS	EMOTIONAL BENEFITS	PRODUCT POSITIONING
Competitor 1	Product Category 1				
	Product Category 2				
Competitor 2	Product Category 1				
	Product Category 2				
Competitor 3	Product Category 1				
	Product Category 2				
Competitor 4	Product Category 1				
	Product Category 2				

Customers attach meanings to brands and develop brand relationships. Brand identity provides the framework for overall brand coherence. It is a concept that serves to offset the limitations of positioning and to monitor the means of expression, unity, and durability of a brand (Kotler 2008).

Brand has a meaning well beyond a product's physical attributes. Brand is a composition of total perceptions in customers' minds: the perception that is made up of the customers' experiences and thoughts, the experiences of trusted people, and messages received from the company.

Brand management is an essential part of future-oriented management. By implementing strong brand management, companies are prepared to adapt to ever-changing competitive conditions (Kotler 2010).

Brand can be considered as an asset like other capital assets owned by a company. *Brand equity* is the term used to describe the value of a brand name. But from the marketer's perspective, brand equity is a tool that can be used by the customers for processing large amounts of information while benchmarking different products or services. Brands are direct consequences of market segmentation and product differentiation strategies (Kotler 2008).

With corporate social responsibility efforts, companies aim to gain credit with the public while strengthening the brand positioning and corporate image and motivating customers and other partners (Kotler and Lee 2005).

Other than corporate social responsibility activities, companies focus on green marketing incentives. Mostly, companies perform green activities by using healthy and renewable materials, renewable energy, and optimized resources (Lannuzzi 2012).

Branding-based competition evaluations are summarized in Table 5.3.

5.3.1.2.3 Promotions-Based Competition Promotions are marketing activities that aim to affect customer purchasing behavior. Promotional activities consist of advertising, public relations, social media, sales promotions, and sales personnel efforts.

5.3.1.2.3.1 Advertising-Based Competition The objective of advertising can simply be described as a specific communication task for a specific target audience during a specific period of time. As it directly transfers the messages from marketers to the targeted audiences, advertising can be accepted as the most reliable marketing tool.

Today, television still makes up the highest proportion of companies' advertorial budgets. In addition to television, newspapers, magazines, radio, outdoor activities, cinemas, brochures, catalogs, direct mailings, blogs, websites, and text messaging are other media tools.

There are three types of advertising: informative, persuasive, and reminder advertising (Ries 2002). Basically, informative advertising focuses on informing the audience about a new brand, a new product, or a price change, or corrects wrong knowledge about the brand or product. Persuasive advertising is mostly used in the periods when the competition is increasing, and the company aims to build brand preference, to

Table 5.3   Branding-Based Competition

	POSITIONING AND BRANDING	CUSTOMER PERCEPTION	BRANDING EFFORTS	CSR AND GREEN MARKETING
Competitor 1				
Competitor 2				
Competitor 3				
Competitor 4				

Note: CSR, corporate social responsibility.

Table 5.4   Advertising-Based Competition

		CONSUMPTION RATE (%) OF MEDIA CHANNELS	CONSUMPTION RATE (%) OF ADVERTISING TYPES	MESSAGES	CUSTOMER/ CONSUMER ORIENTATION	TIMING
Competitor 1	Brand					
	Product Category 1					
	Product Category 2					
Competitor 2	Brand					
	Product Category 1					
	Product Category 2					
Competitor 3	Brand					
	Product Category 1					
	Product Category 2					

affect brand perception, or to persuade the customers to buy. Reminder advertising is mainly used for improving the customer relationship and reminding customers about the brand or products. Advertising-based competition evaluations are summarized in Table 5.4.

5.3.1.2.3.2 Public Relations–Based Competition Public relations (PR) departments perform the following functions:

- Press relations and press agency
- Product publicity
- Public affairs
- Lobbying
- Investor relations

Companies use PR to build good relations with consumers, investors, media, and their communities. Public relations are predominantly used because of their enabling power to create more persuasive impacts on individuals and groups while developing and spreading information about company activities freely to different media channels (Ries 2002). Public relations–based competition is summarized in Table 5.5.

5.3.1.2.3.3 Social Media–Based Competition: Social media is one of the most effective communication tools for today's customers.

Table 5.5 Public Relations-Based Competition

		ACTION	TARGETED AUDIENCE	MEDIA CHANNELS	TARGET PERCEPTION
Competitor 1	Brand				
	Product Category 1				
	Product Category 2				
Competitor 2	Brand				
	Product Category 1				
	Product Category 2				
Competitor 3	Brand				
	Product Category 1				
	Product Category 2				
Competitor 4	Brand				
	Product Category 1				
	Product Category 2				

Even the sales forces are ramping up their use of social networking media from proprietary online customer communities to webinars and even Twitter, Facebook, YouTube and other applications. Social media-based competition is summarized in Table 5.6.

Table 5.6 Social Media Based Competition

		ACTION	TARGETED AUDIENCE	SOCIAL MEDIA CHANNELS	TARGET PERCEPTION/ MESSAGES	INTERACTION RATE OF POTENTIAL CUSTOMERS
Competitor 1	Brand					
	Product Category 1					
	Product Category 2					
Competitor 2	Brand					
	Product Category 1					
	Product Category 2					
Competitor 3	Brand					
	Product Category 1					
	Product Category 2					
Competitor 4	Brand					
	Product Category 1					
	Product Category 2					

5.3.1.2.3.4 Sales Promotions–Based Competition   Promotions that are directly focused at making changes in customers' purchasing decisions are called *sales promotions*. Sales promotions are short-term marketing activities that are used to penetrate the market or promote the sales of a single product, a product segment, or a product category. Price discounts, loyalty bonuses, free shipping services, flash sales promotions, bundles (e.g., buy eight and pay for seven), product giveaways (buy product X and get product Y as a gift), coupon giveaways, price match promises, and lotteries are the most common types of sales promotion.

Any company may aim to penetrate the market with its new brand or product, make the brand or product more memorable, build loyalty to the brand or product, collect data about customers, or clear its stocks by sales promotions.

The main objectives of sales promotions include influencing retailers to carry new items and more inventory, buy ahead, or promote the company's products, and so on (Kotler and Armstrong 2011). Sales promotions–based competition is summarized in Table 5.7.

5.3.1.2.3.5 Sales Personnel–Based Competition   The sales force serves as a critical link between a company and its customers, which can be applied in different forms, such as face-to-face or by telephone calls, video calls, video conferences, or e-mail (Kotler and Armstrong 2011). Recruiting the right people, right training, right comprehension, and right motivation are the basics of improving the quality and efficiency of sales personnel. Sales personnel–based competition is summarized in Table 5.8.

Table 5.7 Sales Promotions–Based Competition

		SALES CHANNELS	PROMOTION MODEL
Competitor 1	Product Category 1		
	Product Category 2		
Competitor 2	Product Category 1		
	Product Category 2		
Competitor 3	Product Category 1		
	Product Category 2		
Competitor 4	Product Category 1		
	Product Category 2		

Table 5.8 Sales Personnel–Based Competition

		QUALIFICATIONS	INITIATIVES
Competitor 1	Sales Channel 1		
	Sales Channel 2		
Competitor 2	Sales Channel 1		
	Sales Channel 2		
Competitor 3	Sales Channel 1		
	Sales Channel 2		
Competitor 4	Sales Channel 1		
	Sales Channel 2		

5.3.1.2.4 Sales Channels–Based Competition Today, most companies try to optimize their sales channels by being in the right place at the right time for meeting their targeted customers. Optimization of sales channels is mostly implemented by inspecting shopping behaviors and habits.

Sales channel design calls for segmenting consumer needs, setting channel objectives, identifying major channel alternatives, and evaluating those alternatives (Kotler and Armstrong 2011).

The offline sales channel, known as the most traditional bridge between the buyer and the seller, still keeps its dominance in sales volumes in numerous markets. However, the fastest-growing sales tool is the internet. The internet offers explosive potential for conducting sales operations and interacting with and serving customers. Sales organizations are now both enhancing their effectiveness and saving time and money by using a host of internet approaches to train sales reps, hold sales meetings, and even conduct live sales meetings with customers (Kotler and Armstrong 2011). The use of both offline and online sales channels in competition is shown in Table 5.9.

Table 5.9 Sales Channels–Based Competition

		SALES CHANNELS	SHARE (%)	PROFITABILITY
Competitor 1	Product Category 1			
	Product Category 2			
Competitor 2	Product Category 1			
	Product Category 2			
Competitor 3	Product Category 1			
	Product Category 2			
Competitor 4	Product Category 1			
	Product Category 2			

5.3.1.2.5 Customer Processing–Based Competition   Customer process-ing includes both primary and secondary activities (Porter 1998). Primary activities add value to the customer interface while customers are experiencing the service, and in the meantime, various processes are supported by secondary activities, mostly known as the *back office applications*, which support the consumer before, during, and after con-sumption. Designing the processes that customers are subjected to is a strategically important decision for companies and other providers that operate in competitive environments. Increasing the effectiveness and quality of customer processing while decreasing its complexity should be considered highly significant for customer purchasing decisions and should be adapted to all strategic departments of the company. Customer processing–based competition is summarized in Table 5.10.

5.3.1.2.6 Packaging-Based Competition   By differentiating packag-ing from that of the competitors, marketers aim to get the attention of customers at first sight (Kotler and Keller 2012). Design-focused dif-ferentiation is mostly done by conceptual design, adding user descrip-tions, promoting information, bundling, or directly applying price tags to the packages.

Another technique that can be used by marketers is creating addi-tional customer value by adding functionality to the packaging mate-rials. Sometimes, even seemingly small packaging improvements can make a big difference, such as changing the targeted market and the way product is used. As an example, carbonated drink in PET bottles and aluminum cans is offered to individuals, whereas PVC bottles are usually targeted to families (Trott 2005). Packaging-based competi-tion is summarized in Table 5.11.

Table 5.10   Customer Processing–Based Competition

		BEFORE	DURING USE	AFTER
Competitor 1	Product Category 1			
	Product Category 2			
Competitor 2	Product Category 1			
	Product Category 2			
Competitor 3	Product Category 1			
	Product Category 2			
Competitor 4	Product Category 1			
	Product Category 2			

Table 5.11 Packaging-Based Competition

		EMOTIONAL	FUNCTIONAL	PROMOTIONAL	INFORMATIVE
Competitor 1	Product Category 1				
	Product Category 2				
Competitor 2	Product Category 1				
	Product Category 2				
Competitor 3	Product Category 1				
	Product Category 2				
Competitor 4	Product Category 1				
	Product Category 2				

5.3.1.2.7 Partnership-Based Competition Each company hosts differences within its managerial abilities, human resources, financial structure, software and hardware resources, experience, and corporate culture. Therefore, creating effective strategic partnerships enables companies to focus on their core businesses while the other business functions are managed by companies that are ready to focus on these areas. Faced with the new level of competition, many companies start sharing their resources and expertise to develop new products, achieve economies of scale, and gain access to new technologies and markets (Trott 2005).

Twenty-first-century companies are now creating partnerships for delivering better customer values and decreasing their own risks. Delegating functions to third-party suppliers, service providers, advertising and PR agents, sales channels, and even customers enables the companies to create more precious customer values (see Table 5.12).

Table 5.12 Partnerships-Based Competition

		PARTNER	DELIVERED VALUE	BUSINESS MODEL
Competitor 1	Brand Partnership			
	Product Category 1			
	Product Category 2			
Competitor 2	Brand Partnership			
	Product Category 1			
	Product Category 2			
Competitor 3	Brand Partnership			
	Product Category 1			
	Product Category 2			
Competitor 4	Brand Partnership			
	Product Category 1			
	Product Category 2			

5.3.1.2.8 Price/Customer Payoff Competition   Assuming that all companies deliver identical value propositions, pricing would be the most efficient business strategy to compete.

Today, the customer has a large influence on the price, so eliminating costs by being more efficient is a necessity (Wheat 2003). So, reducing costs with innovative solutions should be considered as *pricing innovation*. Other than cost reduction, the pricing model can also be viewed as an effective marketing tool (see Table 5.13).

5.3.2 Investigate and Clarify the Market Gap

Our first fundamental is to investigate the market and understand the market gap. Understanding the market gap or the market opportunity by using several analytical techniques would be necessary to state the upcoming trends of customer purchasing behaviors. While trying to find out the market gap, different ideas and solution offerings would be submitted. Choosing the most promising market gap and the customer trend for the business would finalize this step.

Customer-based new product development focuses on finding new ways to solve customer problems and create more customer-satisfying solutions. Thus, customer involvement has a positive effect on the new product development process (Kotler and Armstrong 2011).

Therefore, for better understanding of the next generation of customer trends, the evolution of each marketing mix should be analyzed. In this step, planners investigate the evolutions of the marketing mix components with respect to the targeted customers' expectations. *Basic* expectations refer to the essential basics that are needed to compete in the market. *Satisfactory* expectations refer to additional

Table 5.13   Price/Customer Payoff–Based Competition

		COST REDUCTION MODEL	PRICING MODEL
Competitor 1	Product Category 1		
	Product Category 2		
Competitor 2	Product Category 1		
	Product Category 2		
Competitor 3	Product Category 1		
	Product Category 2		
Competitor 4	Product Category 1		
	Product Category 2		

positive values to the customer. The *Unexpected* is the innovative one, which is considered unexpected by customers. Unexpected expectations refer to additional positive values that go beyond the satisfactory expectations of customers.

For example, at the beginning, the anti-lock braking system (ABS) in vehicles was an unexpected improvement for all users. The majority of customers perceived ABS as unexpected for a long time. After a while, customers began to perceive it as satisfactory, and finally, today, ABS is accepted as basic.

In this step, planners should conduct brainstorming sessions about the following preidentified marketing mix topics. During the brainstorming sessions, the competition analysis section's outcome should be used, and the results of this step should be transferred to the relevant tables in this chapter to position the current conditions of the market.

The brainstorming sessions should be conducted by employees from different departments, professionals from different disciplines, academicians, authorities, and customers.

For the following subsections, use the tables for the corresponding subsections in Section 5.3.1. In that respect, use the same columns as the corresponding tables, replace their rows with *unexpected*, *satisfactory*, and *basic* headings (as shown in Table 5.14), and for each table, do the necessary evaluations accordingly.

5.3.2.1 Evolution of Product and Service In this section, planners investigate the next-generation product trends (see Table 5.14).

5.3.2.2 Evolution of Branding Here, planners investigate the next-generation customer trends within the context of branding innovations.

Table 5.14 Evolution of Product and Service

	PHYSICAL	FUNCTIONAL BENEFITS	EMOTIONAL BENEFITS	PRODUCT POSITIONING
Unexpected				
Satisfactory				
Basic				

They also investigate the next-generation customer trends within the context of corporate social responsibility and green marketing innovations.

5.3.2.3 Evolution of Promotions   The evolution of promotional activities consists of advertising, public relations, social media, sales promotions, and sales personnel.

5.3.2.3.1 Evolution of Advertising   Here, planners investigate the next-generation customer trends within the context of advertising innovations. A successful innovation has the ability to create its own advertising without additional expense. Accepted as the most efficient tool of advertising, "word of mouth" creates the highest rates of return on customer behaviors and, in parallel, on sales volumes, while innovation also enables visibility on various media tools.

5.3.2.3.2 Evolution of Public Relations   Here, planners investigate the next-generation customer trends within the context of public relations innovations with respect to the action, the targeted audience, media channels, and the target perception.

5.3.2.3.3 Evolution of Social Media   Social media can be accepted as the most interactive platform that enables companies to keep in touch with their consumers.

5.3.2.3.4 Evolution of Sales Promotions   Sales promotions have been the most used promotions tool for years. The promotion model refers to the promotions tool or the sources of the promotions tool, while this model is observed in each sales channel.

5.3.2.3.5 Evolution of Sales Personnel   As sales personnel enable face-to-face interaction, they have always taken an important role since the beginning of marketing. The efforts of the sales personnel should be investigated with respect to their qualifications and initiative-taking capabilities.

5.3.2.4 Evolution of Sales Channels   In this section, planners investigate the next-generation customer trends within the context of sales channels innovations. The evolution figures of the sales channels for a defined market should be investigated by considering the evolution of market shares and the profitability of related sales channels.

5.3.2.5 Evolution of Customer Processing Customer processing is becoming less complicated and easier as customers choose companies that undertake these responsibilities on their behalf. This section investigates the before-, during-, and after-processing evolution.

5.3.2.6 Evolution of Packaging Packaging includes functional, promotional, and informative aspects in practice. Planners investigate the next-generation customer trends within the context of packaging.

5.3.2.7 Evolution of Partnerships Today, most companies accept partnerships as an opportunity to focus on their core businesses while delegating the operations, or even the management, of multiple functions.

5.3.2.8 Evolution of Price/Customer Payoff Here, innovative cost-reducing and pricing models are considered. Planners investigate the next-generation customer trends within the context of price/customer payoff.

5.3.3 Clarify Capabilities and Relate to Market Gap

When problems and opportunities are defined as *market gap*, it is necessary to conduct a preliminary assessment on the capabilities (resources and abilities) of the company and its networking relationships to understand whether feasible solutions exist prior to assigning substantial resources to filling the determined market gap. During the assessment, analysts usually work with the representatives of the relevant department(s) that expect to benefit from the outcomes (Overton 2007).

The required capabilities, such as *financial, managerial, human resources, sales and marketing abilities, software and hardware capabilities, research and development capabilities, know-how, networking, and corporate culture*, should be analyzed to understand whether the company is able to fill the determined market gaps.

For the following subsections, use the tables for the corresponding subsections in Section 5.3.1. In that respect, use the same columns as the corresponding tables, replace their rows with *own capabilities* and *third-party capabilities* headings (as shown in Table 5.15), and for each table, do the necessary evaluations accordingly.

5.3.3.1 Product and Service Providing–Related Capabilities   In this section, the branding capabilities of the company are analyzed to understand whether feasible solutions exist for filling the determined market gap (see Table 5.15). As mentioned, the company may handle the requirements with its own resources and/or by building partnerships with third parties.

5.3.3.2 Branding-Related Capabilities   In this section, the branding capabilities of the company are analyzed to understand whether feasible solutions exist for filling the determined market gap.

5.3.3.3 Promotions-Related Capabilities
Advertising-related capabilities: Here, the advertising capabilities of the company are analyzed to understand whether feasible solutions exist for filling the determined market gap.
   Public relations–related capabilities: Here, the public relations capabilities of the company are analyzed to understand whether feasible solutions exist for filling the determined market gap.
   Social media usage–related capabilities: Here, the social media usage–related capabilities of the company are analyzed to understand whether feasible solutions exist for filling the determined market gap.
   Sales promotions–related capabilities: Here, the sales promotions capabilities of the company are analyzed to understand whether feasible solutions exist for filling the determined market gap.

Table 5.15   Product and Service Providing–Related Capabilities

	PHYSICAL	FUNCTIONAL BENEFITS	EMOTIONAL BENEFITS	PRODUCT POSITIONING
Own Capabilities				
Third-party Capabilities				

Sales personnel–Related capabilities: Here, the sales personnel capabilities of the company are analyzed to understand whether feasible solutions exist for filling the determined market gap.

5.3.3.4 Sales Channels–Related Capabilities   In this section, the sales channels capabilities of the company are analyzed to understand whether feasible solutions exist for filling the determined market gap.

5.3.3.5 Customer Processing–Related Capabilities   In this section, the customer processing capabilities of the company are analyzed to understand whether feasible solutions exist for filling the determined market gap.

5.3.3.6 Packaging-Related Capabilities   In this section, the packaging capabilities of the company are analyzed to understand whether feasible solutions exist for filling the determined market gap.

5.3.3.7 Partnerships-Related Capabilities   In this section, the partnerships capabilities of the company are analyzed to understand whether feasible solutions exist for filling the determined market gap.

5.3.3.8 Pricing/Customer Payoff–Related Capabilities   In this section, the pricing capabilities of the company are analyzed to understand whether feasible solutions exist for filling the determined market gap.

5.3.4 Develop Alternative Innovations

The innovation-generation process includes idea generation, idea screening, concept testing, marketability analysis, product/service development, commercialization testing, monitoring, and evaluation. Ideas are generated and screened by brainstorming sessions.

In this step, the planners analyze the marketability of the generated and screened innovation ideas, while the design and manufacturability should be applicable for potential innovations.

Assuming that *market size* and *market growth rates* are in acceptable ranges, the analysis described in this and the following paragraph should be applied for defining the marketability of the potential innovations. *Readiness* refers to the expectation rate and the infrastructure readiness (hardware and software) of the targeted customer segment. The *satisfaction rate* may refer to the total satisfaction or the marginal satisfaction rates through the use of innovated customer values. After defining these details and the *price sensitivity* of the targeted customer segment, the *sales* and the *profitability forecastings* are defined. The *project duration* refers to the duration of the implementation, and the *cost* refers to the total cost including the human resources.

For the following subsections, use the tables for the corresponding subsections of Section 5.3.1. In that respect, use the same columns as the corresponding tables, replace their rows with the row headings of Table 5.16, as shown on the following page, and for each table, do the necessary evaluations accordingly.

5.3.4.1 Marketability Analysis of Potential Product and Service Innovations In this section, the marketability of defined product/service innovations is analyzed with respect to the readiness of customers for innovative solutions (see Table 5.16).

5.3.4.2 Marketability Analysis of Potential Branding Innovations In this section, the marketability of defined branding innovations is analyzed with respect to the readiness of customers for innovative solutions. As branding is a multiple-step function, the company may undertake numerous efforts to generate the desired brand positioning and may cooperate with third parties for this purpose.

5.3.4.3 Marketability Analysis of Potential Promotion Innovations Advertising innovations: Here, the marketability of defined advertising innovations is analyzed with respect to the readiness of customers for innovative solutions.

Public relations–related innovations: Here, the marketability of public relations–related innovations is analyzed with respect to the readiness of customers for innovative solutions.

Social media–related innovations: Here, the marketability of defined social media–related innovations is analyzed with respect to the readiness of customers for innovative solutions.

Sales promotions–related innovations: Here, the marketability of sales promotions–related innovations is analyzed with respect to the readiness of customers for innovative solutions.

Table 5.16 Marketability Analysis of Potential Product and Service Innovations

	PHYSICAL		FUNCTIONAL BENEFITS		EMOTIONAL BENEFITS		PRODUCT POSITIONING	
	INNOVATION 1	INNOVATION 2	INNOVATION 1	INNOVATION 2	INNOVATION 1	INNOVATION 2	INNOVATION 1	INNOVATION 2
Customer technical readiness for innovation (%)								
Necessity for customer (%)								
Customer willingness for marginal cost (%)								
Expected satisfaction rate (%)								
Benefits on brand (%)								
Cost of innovation								
Sales forecasting								
Profitability forecasting								
Project duration								
Expected market share in maturity (%)								
Patent protection ability								

Sales personnel–related innovations: Here, the marketability of defined sales personnel–related innovations is analyzed with respect to the readiness of customers for innovative solutions.

5.3.4.4 Marketability Analysis of Potential Sales Channels Innovations   In this section, the marketability of defined sales channels innovations is analyzed with respect to the readiness of customers for innovative solutions.

5.3.4.5 Marketability Analysis of Potential Customer Processing–Related Innovations   In this section, the marketability of defined customer processing–related innovations is analyzed with respect to the readiness of customers for innovative solutions.

5.3.4.6 Marketability Analysis of Potential Packaging Innovations   In this section, the marketability of defined packaging innovations is analyzed with respect to the readiness of customers for innovative solutions.

5.3.4.7 Marketability Analysis of Potential Partnerships-Related Innovations   In this section, the marketability of defined partnerships-related innovations is analyzed with respect to the readiness of customers for innovative solutions.

5.3.4.8 Marketability Analysis of Potential Price/Customer Payoff–Related Innovations   In this section, the marketability of defined pricing-related innovations is analyzed with respect to the readiness of customers for innovative solutions.

5.3.5 Evaluate and Accept the Best Plan

Preliminary screening for alternative marketing innovation plans should be separately evaluated, and the best applicable one should be selected using the Evaluation of Alternatives Worksheet.

Identify each alternative appropriately as described in Step 4. Prepare a worksheet showing the comparable costs of each alternative. The marketing team should work together for estimating the most reliable costs of each alternative.

Also, on a separate worksheet, make a comparison of the intangible benefits and risks of each alternative. Compare alternatives, select the best, and get the others to approve it. Enter the headings on a fresh copy of the worksheet (see Figure 5.2), generated by Muther (2011), checking the box marked *intangibles* (upper left). Identify each alternative by a letter—X, Y, Z–and give a brief two-to-five-word description of each.

List all factors, considerations, or objectives the organization wants the project's intended plan to achieve:

- Customers' technical readiness for innovation (%)
- Necessity of innovation for customers (%)
- Expected satisfaction rate (%)
- Customer willingness for marginal cost (%)
- Benefits on brand (%)
- Cost of innovation ($)
- Sales forecasting ($)
- Profitability forecasting ($)
- Project duration
- Expected market share in maturity (%)
- Patent protection ability

Then ask them to weigh the importance of each other factor relative to the most important (10). Indicate each selected weight on the worksheet, and record by whom the weight values were determined.

Ask your operations team and/or staff members who will use the proposed plan when installed to rate, for each factor, the effectiveness of each alternative in achieving that factor. Use A, E, I, O, or U to represent descending order of effectiveness, as noted in the upper left-hand box of the worksheet. Enter the selected vowel-letter ratings in the small rectangular *boxes within boxes* on the form. Record the name(s) of the person(s) doing the rating.

After rating all alternatives for each factor, convert letters to numbers (A = 4, E = 3, I = 2, O = 1, U = 0) and multiply the rated number by the respective weight value. Enter the resulting weighted-rated values on the worksheet.

Down-total the weighted-rated values for each alternative, enter into the worksheet, and record by whom the tally was made. The

Evaluating alternatives

Costs
☐
Estimated by _____ Approved by _____
Total Annualized Cost = Investment Cost _____ +Annual Operating Cost
Expected Life

Intangibles
☐
Weight set by _____ Tall by _____
Ratings by _____ Approved by _____
Evaluating description

A	Almost perfect	O	Ordinary results
E	Especially good	U	Unimportant results
I	Important results	X	Not acceptable

Project _____ Number _____
By _____ With _____
Date Sheet _____

Description of alternatives

X. _____
Y. _____
Z. _____
V. _____
W._____

Factor/consideration	WT	Alternative				
		X	Y	Z	V	W
1						
2						
3						
4						
5						
6						
7						
8						
9						
10						
11						
12						
13						
14						
15						

Total
☐ Annualized cost (line_____ Plus line _____)
☐ Weighted rated down total

Reference notes:
a. _____ d. _____
b. _____ e. _____
c. _____ f. _____

Figure 5.2 Evaluation of alternatives worksheet. (From Muther R., *Planning by Design*, Institute for High Performance Planners, Kansas City, 2011.)

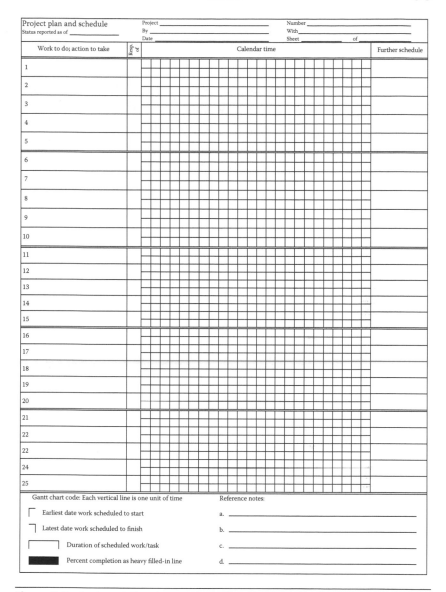

Figure 5.3 Implementation plan worksheet. (From Muther R., *Planning by Design*, Institute for High Performance Planners, Kansas City, 2011.)

alternative with the highest total should be the *winner*—subject to cost factors determined separately. In the lower left corner, indicate that these are weighted-rated down totals. Record any explanatory notations at the bottom, suitably referenced by an encircled lower-case letter.

5.3.6 *Implementation Plan*

This step is dedicated to carrying out the selected innovation in terms of its development, commercialization testing, monitoring, and evaluation. In implementation, who does what is very important. It includes the actions needed to make the plan come true. The person(s) responsible and the duration of action have to be clarified. This step sets the framework for dealing with the "expectations" on time. We can use the MS Project tool in this step (see Figure 5.3).

6

SYSTEMATIC COMPETENCY-BASED TRAINING NEEDS ANALYSIS (SCTNA)

HAKAN BÜTÜNER AND DIDEM HACIPASAOGLU

6.1 Introduction to SCTNA

There are some features that employees must have for accomplishing the requirements of their jobs. Competency-based training needs analysis deals with the identification of the training that must be undertaken by the employees based on the comparison of the required features and their existing competencies.

This project should not only upgrade the suitability of the employees' behaviors to the organization but should also help reach the goals of the organization. By paying attention to the existing jobs, values and career paths of individuals, it is possible to upgrade the employees' knowledge, abilities and competencies and provide the necessary behavioral change. This working model outlines a systematic methodology for competency-based training needs analysis.

Why do we need training?

- Increasing competencies
- Increasing performance
- Understanding the strategies and goals of the company
- Meeting the expectations of the customers
- Developing new products
- Developing career plans
- Redesigning the job content

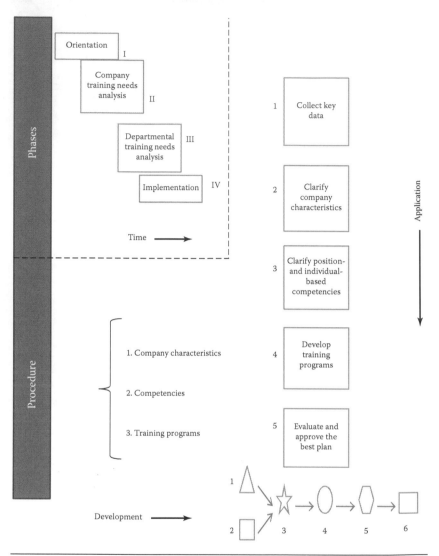

Figure 6.1   SCTNA: Reference sheet.

This working model outlines a systematic methodology for competency-based training needs analysis (see Figure 6.1).

6.2 Four Phases

Any competency-based training needs analysis can typically pass through four phases.

6.2.1 Orientation

The main question of the orientation phase is "What is our objective?" in this project. This means to orient ourselves and to understand the project, the process, and the people involved. Then, we organize how we propose to plan and schedule the planning. The main issue is "What do we do?" and "How do we do it?"

- Understand the project: What? Why? Who? When? Where?
- Understand the purpose or objective(s), the external conditions, the situation(s), the scope/extent, any budget limitations, and the desired form of our planning output
- Understand and document the planning and people issues
- Make a schedule for the project planning

We can use the "Project Orientation Worksheet" designed by Muther (2011) (Table 6.1). This has three components:

- Project essentials
- Planning issues
- Planning schedule

In "Project Essentials," enter

- The objective(s) or purpose(s) or goal(s) of this project
- The external conditions, such as synchronization with other projects, specific limitations, overall policies, or larger operational procedures …
- The situation(s): Physical, procedural, and personal situation circumstances
- The scope/extent of the project: How big? How detailed? When needed?
- The form of this planning output: Written report? Action plan approved?

Goals for individuals:

- Increasing their motivation
- Helping them to reach their targeted careers
- Creating harmony and collaboration among them
- Providing satisfaction
- Improving their competencies

Table 6.1 Project Orientation Worksheet

Description ———————————————— Project no:—————		

Who is responsible? ———————— Authorized/Initiated by ————Date————

When project starts ———————— When planning starts ———— Sheet of —

Project essentials

1. Project objective(s) ———————————————————
2. External condition(s) ———————————————————
3. Situation(s) ———————————————————
4. Scope/extent ———————————————————
5. Form of output ———————————————————

Planning issues	Imp.	Resp.	Proposed resolution	Ok'd by
1.				
2.				
3.				
4.				

Planning schedule										Notes and act
Task or action required to plan	Who									
1.										
2.										
3.										
4.										

Reference notes: ———————————————————

Source: Muther, R., *Planning by Design*, Institute for High Performance Planners, Kansas City, MO, 2011.

Goals for the company:

- Increasing efficiency and effectiveness
- Helping people to adjust more easily to changes
- Decreasing the error rate

In "Planning Issues," we enter each problem, uncertainty, and question—one line for each—on the left. In the first column, record how important the issue is to this project. Here, we enter a vowel-letter as our order-of-magnitude judgmental rating:

A: Absolutely important
E: Especially important
I: Important issue
O: Ordinary important
U: Unimportant

In the "Responsible" column, enter who is responsible for getting the issue resolved, and add the initials of the approver.

In "Planning Schedule," list each action required to plan what we intend to do to prepare training needs analysis for conducting the project. List one action on each line, and show who is responsible for doing it. Set a calendar schedule at the top of the vertical lines.

6.2.2 Overall (Company) Training Needs Analysis

The general requirements and principles of company-wide training will be determined. The aim in this phase is to determine the overall competency-based training needs, in such a way that in principle, they meet the objectives of the company and integrate with the external conditions.

6.2.3 Detailed (Departmental) Training Needs Analysis

Detailed competency-based training needs analysis will be applied to understand the requirements of every individual function (department), which in turn, means that three fundamentals are valid for every department. In short, this is a competency-based training needs analysis conducted for each department, as identified in Phase 1. Phase 3 repeats the same essential analysis process, but it does so at a more specific level.

6.2.4 Implementation

This is probably the most rewarding phase. This phase is dedicated to carrying out the plans. It includes the actions needed to make the plans come true. Who will be responsible and the duration of action have to be clarified. This phase sets the framework for dealing with the *expectations* on time and on budget. We can use the MS Project tool in this phase. Table 6.2 shows the output of the project that is

Table 6.2 Sample Implementation Plan Worksheet

Actions to take	Team	Budget	Week 1	Week 2	Week 3	Week 4	Week 5	Week 6	Week 7	Week 8	Week 9	Week 10	Week 11	Week 12
Company polities and strategies			▓	▓										
Job analysis and job descriptions				▓										
Identification of the competencies					▓									
Preparing the positional surveys						▓								
Sending and taking back the positional surveys							▓							
Analyzing the positional surveys								▓						
Integration of the competencies and trainee program									▓	▓				
Preparing the individual surveys										▓				
Sending and taking back the individual surveys											▓	▓		
Analyzing of the individual surveys													▓	
Outputs of the project														▓

prepared for each trainee group. One of the key points is to identify and underline the expected competencies from each training program.

6.3 Three Fundamentals

The three fundamentals of SCTNA are company (department) characteristics, competencies, and training programs (see Figure 6.2).

6.3.1 Company (Department) Characteristics

Based on the strategies and values of the company (department), identifying the requirements for structural positions is considered as Fundamental A. Accordingly, the questions that should be investigated are

- What are the causes of organizational weaknesses?
- Is training a necessity?
- What degree of training is needed?
- What should be the aim of the training?

6.3.2 Position- and Individual-Based Competencies

In position-based competencies, by referring to the job specifications, the competencies that are necessary for each position are determined.

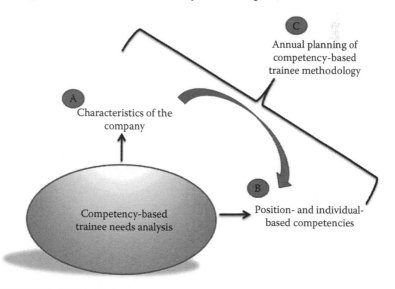

Figure 6.2   SCTNA: fundamentals.

In individual-based competencies, personal gaps and therefore personal competencies that are necessary for meeting the required specifications are determined. At the end, both of the results should be integrated.

6.3.3 Training Programs

Fundamental C, which is formed by the integration of the first and second fundamentals, consists of the training programs specific for each department and generic for the whole organization. Budget plans, time plans, and contents of the training are all covered within the scope of this fundamental.

6.4 Five Sections

In the framework of the full planning pattern, for the second and third phases, you need to pass through the following five sections to develop the best possible competency-based training programs.

The short form condenses the four phases into six steps and combines Phase 2 and Phase 3. The short form is applicable to short or smaller planning assignments or situations.

6.4.1 Collect Key Data

There is some significant information that must be investigated at the beginning of the project. These investigations are categorized into two types. The first should be done with the top-level management of the company to understand the vision, mission, strategies, and values of the company.

The second should be done with the human resources department. The second investigation includes the job analysis and descriptions of the organizational structure.

6.4.2 Investigate and Clarify Company (Department) Characteristics

Clarifying company characteristics starts with investigating vision, mission, basic values, and strategies, and continues with understanding organization structure and job descriptions. In short, characteristics of the company are categorized as

- General information about the company
- Job descriptions and organizational scheme
- Competency model

The competency model (Naquin and Holton 2006) is the determination of the managerial and leadership competencies, in addition to the basic and technical competencies that are specifically necessary for the firm. Identification of the competencies is done through the understanding of job analysis, descriptions, and organizational scheme. Every main competency in the competency model has subtitles that should be determined.

These competencies are necessary for all employees, independently of their positions and titles. These competencies can be different according to the company (department) and its industry.

Managerial competencies are understood by investigating the job analysis and job descriptions. These competencies are determined through the identification of middle- and top-level managers' requirements. These competencies should be unique to the company (department).

Leadership competencies are necessary for the leadership positions, such as team leaders, of the company (department). They should also have basic competencies, such as effective communication, to be an effective leader for their team.

Although technical competencies are a *must* for all the employees in the company, they can change rapidly from level to level or from department to department. While the marketing department requires information about the product and the market, the product department requires information about process/process management.

Significant points that must be considered while clarifying the competencies are

- Focusing on the main facilitating area of the company
- Regarding the organization, not individuals, while identifying basic competencies
- Not making a sharp discrimination while clarifying the leadership, managerial, and technical competencies

6.4.3 Clarify Position- and Individual-Based Competencies and Relate to Company (Department) Characteristics

The human resources department decides to what degree competencies are necessary for the respective positions. This is done for every position.

Individual competencies are determined by a survey that is applied to all employees. The aim of this survey is to identify the required competency levels for the employees. Based on the strengths and weaknesses of the employees, the competency level gaps are determined.

After determining the subcompetencies in the second step, we develop the behavioral indications of these competencies, which help us establish the content of training programs. An example form is given in Table 6.3.

A positional survey form (see Table 6.4) is generated with the help of the organizational scheme and applied to every position.

A position and competency matrix (see Table 6.5) integrates the positions and the competencies. The aim is to understand the required competency level for each position.

The survey determines the required level of competencies for each position:

A: Absolutely necessary
E: Especially necessary
I: Important
O: Ordinary
N: Not necessary

Individual surveys can also be named the *individual-based competency inventory*. After identifying the competencies for each position,

Table 6.3   Example Competency Form

Effective communication		
Effective communication is an essential component of organizational success whether it is at the interpersonal, intergroup, intragroup		
		Behavioral indications
Subcompetencies	Oral communication	Effective and clear statement of what is desired to be said ...
	Written communication	....
	Body language	....
	Empathy	....

find out the individuals' capabilities and their gaps with respect to the required competencies by investigating their strengths and weaknesses.

6.4.4 Develop Training Programs

The main purpose of this step is to fill the required competency levels through deploying the necessary training. Based on the detected weaknesses of the employees due to the required competency levels, annual competency-based training plans are set.

After the training programs have been matched to the desired competencies, the time plans, budgets, and contents of the programs are determined.

Table 6.4 Positional Survey Form

Position: Marketing supervisor		N	O	I	E	A
A: Absolutely need E: Especially need I: Importantly need O: Ordinarily need N: Not necessary need						
Effective communication is an essential component of organizational success whether it is at the interpersonal, intergroup, intragroup...	1	X				
Effective and clear statement of what is desired to be said...	2		X			
....	3			X		
....	4					
....	5					
....	6					
....	7					
....	8					
....	9					
	10					
	11					
	12					
	13					
	14					
	15					
	16					

Every competency requires a different training program. At the end of the individual and positional analysis, training plans are developed for each department and employee (see Table 6.6).

The last part of this section is concerned with forming the contents of training programs. A form that can be used for each training title is shown in Table 6.7.

6.4.5 Evaluate and Accept the Best Plan

Identify each alternative appropriately, as described in Section 4. Prepare a worksheet showing the comparable costs of each alternative. Also, on a separate worksheet, make a comparison of the intangible benefits and risks of each alternative. Compare alternatives, select the best, and get the others to approve it.

6.4.5.1 Comparison Based on Intangibles   Enter the headings on a fresh copy of the worksheet (see Table 6.8) generated by Muther (2011),

Table 6.5   Position and Competency Matrix

Classification of the competency		Marketing department			
Main competency					
Behavioral indication of the main competency		Marketing manager	Marketing team leader	Marketing specialist	Marketing assistant
Subcompetencies	Behavioral indications				
....	....				
....	....				
....	....				
....	....				
Basic competencies		Marketing department			
Effective communication					
Effective communication is an essential component of organizational success whether it is at the interpersonal, intergroup, intragroup. ...		Marketing manager	Marketing team leader	Marketing specialist	Marketing assistant
Subcompetencies	Behavioral indications				
Oral communication	Effective and clear statement of what is desired to be said...	A	I	E	O
Written communication	....	E	O	E	A
Body Language	....	A	A	A	I
Empathy	....	A	E	O	N

Table 6.6 Training Plans

Trainings	Time period of the training	Attendance	Number of groups	Total period of the training	Cost of the training	Competencies	Timing
Supervisor group							
Management training	2 Full day	40	2	4 Days	4000 USD	*Planning *Controlling	17–18 August 2008
				4 Days	4000 USD		
Specialist group							
Effective communication training	2 Full day	15	1	2 Days	2000 USD	*Verbal communication *Written communication	20–21 June 2008
				2 Days	2000 USD		

Table 6.7   Detailed Training Programs

Program plan			
Name of the program			
Code of the program			
Timing			
Trainers			
Kind of program			
Methodology of the program			
Quota of the program			
Targets of the program	Knowledge/Ability Behaviors Job results		
Assessment of the effectiveness of the program			
During the program	Who	When	How
Knowledge/Ability			
Behaviors			
After the program	Who	When	How
Knowledge/Ability			
Behaviors			
Job results			

checking the box marked *intangibles* (upper left). Identify each alternative by a letter—X, Y, Z—and give a brief two-to-five-word description of each.

List all factors, considerations, or objectives the organization wants the project's intended plan to achieve. Select, or ask your approvers to select, the most important factors. Then, ask them to weigh the importance of each factor relative to the most important (10). Indicate each selected weight on the worksheet, and record by whom the weight values were determined.

Ask your operations team and/or staff members who will use the proposed plan when installed to rate, for each factor, the effectiveness of each alternative in achieving that factor. Use A, E, I, O, or U to represent the descending order of effectiveness, as noted in the upper

Table 6.8 Evaluation of Alternatives Worksheet

Evaluating alternatives		Project _____ Number _____
☐ Costs:		By _____ With _____
Estimated by .... Approved by		Date _____ Sheet _____ of ___

☐ Intangibles:	
Weight set by .... Tally by	Description of alternatives:
Ratings by .... Approved by	X. _____
Evaluating description	Y. _____
A = Almost perfect, O = Ordinary pesult	Z. _____
E = Especially good, U = Unimportant results	V. _____
I = Important result, X = Not acceptable	W. _____

	Factor/consideration	WT.	Alternative				
			X	Y	Z	V	W
1.							
2.							
3.							
4.							
5.							
6.							
7.							
8.							
9.							
10.							
11.							
12.							
13.							
14.							
15.							
Total	Annualized cost (line____plus line____)						
	Weighted rated down total						

Reference notes:

a. _____   d. _____

b. _____   e. _____

c. _____   f. _____

Source: Muther, R., *Planning by Design*, Institute for High Performance Planners, Kansas City, MO, 2011.

left-hand box of the worksheet. Enter, in the small rectangular *boxes within boxes* on the form, the selected vowel-letter ratings. Record the name(s) of the person(s) doing the rating.

After rating all alternatives for each factor, convert letters to numbers (A = 4, E = 3, I = 2, O = 1, U = 0) and multiply the rated number by the respective weight value. Enter the resulting weighted-rated values on the worksheet.

Down-total the weighted-rated values for each alternative, enter into the worksheet, and record by whom the tally was made. The alternative with the highest total should be the *winner*—subject to cost factors determined separately. In the lower left corner, indicate that these are weighted-rated down totals. Record any explanatory notations at the bottom, suitably referenced by an encircled lower-case letter.

6.4.5.2 Comparison Based on Costs   In addition to the intangibles, the cost of the programs is also important. Identify each alternative by a letter—A, B, C—and give a brief three-to-five-word description of each. This comparison will generally let you identify the lowest-cost alternative.

Table 6.9   Departmental Attendance of Employees

TRAINEES/ DEPARTMENTS	MARKETING DEPARTMENT	IMPORT AND EXPORT DEPARTMENT	CALL CENTER	TOTAL
Effective communication trainee	10	30		40
Written communication trainee	25	27		52
Body language trainee	17	47		64

Table 6.10   Example Budget Form

TRAINING PROGRAM	PERIOD OF THE TRAINEE	COST OF THE TRAINING ($)	NUMBER OF GROUPS	TOTAL PERIOD OF THE TRAINING PROGRAM	TOTAL COST ($)
Effective communication trainee	2 days	2000	2 groups	4 days	4000
Written communication trainee	1 day	1000	2 groups	2 days	2000
Body language trainee					

As a result of Section 4, departmental attendance of employees for the training courses was delivered (see Table 6.9).

Once the table is formed, the groupings are done. For example, 50 people may want to attend the effective communication training. Fifty needs to be divided into groups according to the defined group capacities. This is necessary for the timing and budgeting of training. A budget example is given in Table 6.10.

7

Systematic Wide Area Network Planning (SWANP)

HAKAN BÜTÜNER, SINEM AYDOĞDU, AND DOĞAN UÇAR

7.1 Introduction to SWANP

For businesses with multiple locations, remote or home users, or integrated communications with specific vendors or partners, the Wide Area Network (WAN) is their lifeblood. Information technology (IT) organizations face pressure to increase WAN productivity, improve application performance, support global collaboration, improve data protection, and minimize costs. Integrating today's effective WANs beyond traditional capabilities while improving performance and minimizing costs is crucial.

The main purpose of a WAN is to provide reliable, fast, and safe communication between two or more places (nodes) at affordable prices. WANs enable an organization to have a single network connecting all of its departments and offices, even if they are not all in the same building, city, or even continent. In the increasingly globalized marketplace, WANs have become an integral element of many businesses' networks.

Our aim in this systematic methodology is to provide planning tools for network engineers. This working model outlines a systematic methodology for enterprise WAN design (see Figure 7.1).

7.2 Four Phases

Any network plan can typically pass through four phases.

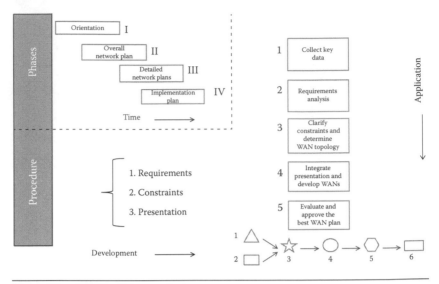

Figure 7.1   SWANP—Reference sheet.

7.2.1 Orientation

The main question of orientation is "What is our objective?" in this project. This means to orient ourselves and to understand the project, the process, and the people involved. Then, we organize how we propose to plan and schedule the planning. The main issue is "What we do?" and "How do we do it?"

- Understand the project: What? Why? Who? When? Where?
- Understand the purpose or objective(s), the external conditions, the situation(s), the scope/extent, any budget limitations, and the desired form of our planning output
- Understand and document the planning and people issues
- Make a schedule for the project planning

We can use the "Project Orientation Worksheet" designed by Muther (2011) (Table 7.1). This has three components:

- Project essentials
- Planning issues
- Planning schedule

Table 7.1   Project Orientation Worksheet

Description ——————————————————— Project no:_____

Who is responsible?   _____   Authorized/Initiated by _____ Date _____

When project starts   ————————— When planning starts   ————— Sheet of —

Project essentials

1. Project objective(s)   ———————————————————————
2. External condition(s) ———————————————————————
3. Situation(s)   ———————————————————————
4. Scope/extent   ———————————————————————
5. Form of output   ———————————————————————

Planning issues	Imp.	Resp.	Proposed resolution	Ok'd by
1.				
2.				
3.				
4.				

Planning schedule							Notes and act
Task or action required to plan	Who						
1.							
2.							
3.							
4.							

Reference notes: ———————————————————————

Source: Muther, R., *Planning by Design*, Institute for High Performance Planners, Kansas City, 2011.

In "Project Essentials," enter

- The objective(s) or purpose(s) or goal(s) of this project
- The external conditions, such as synchronization with other projects, specific limitations, overall policies, or larger operational procedures …
- The situation(s): Physical, procedural, and personal situation circumstances
- The scope/extent of the project: How big? How detailed? When needed?
- The form of this planning output: Written report? Action plan approved?

In "Planning Issues," we enter each problem, uncertainty, and question—one line for each —on the left. In the first column, record how important the issue is to this project. Here, we enter a vowel-letter as our order-of-magnitude judgmental rating:

A: Absolutely important
E: Especially important
I: Important issue
O: Ordinary important
U: Unimportant

In the "Responsible" column, enter who is responsible for getting the issue resolved, and add the initials of the approver.

In "Planning Schedule," list each action required to plan what we intend to do to prepare network plans for conducting the project. List one action on each line, and show who is responsible for doing it. Set a calendar schedule at the top of the vertical lines.

7.2.2 Overall Network Plan

The aim in this phase is to determine an overall plan in such a way that in principle, it meets the objectives of the project and integrates with the external conditions.

Thus, Phase 2 is the process of converting the tangible requirements and situational considerations into a proposed plan that, on the whole, will meet the objectives. This is a plan for the whole situation or total system, as identified in Phase 1.

The overall network plan is larger or more comprehensive than the detailed network plans. Moving from overall plan to detail plans is typified by expressions such as

- From the whole to the parts
- From system to subsystems
- From general to specifics
- From principle to practice
- From policy to procedures

7.2.3 Detailed Network Plans

Phase 3 repeats the same essential planning process, but it does so at a more specific level. Note that the output of Phase 2 is for detailed network plans. Phase 2 planning, for example, might be a connection of locations for the X Company. Within that, Phase 3 planning would involve plans for available network technologies, for clarification of required application and services, and for WAN device specifications. Characteristics that distinguish Phase 3 from Phase 2 include all or many of the following:

- More details and more specifics
- Several plans (compared with only one in Phase 2)
- Smaller spaces or areas considered
- Data or input better understood
- Planners with more particular and less comprehensive skills
- Planners more junior in the organization
- More man-hours required for planning

7.2.4 Implementation Plan

This is probably the most rewarding part of planning. This phase is dedicated to carrying out the plans. It includes the actions needed to make the plans come true. Who will be responsible and the duration of action has to be clarified. This phase sets the framework for dealing with the *expectations* on time and on budget. We can use the MS Project tool in this phase (see Table 7.2).

It is very common for the planners to turn the project over to the implementers. The point is that a planner should be involved in Phase 4, either directly and completely or, at least, partially.

7.3 Three Fundamentals

The three fundamentals of IT WAN planning are requirements, constraints, and presentation.

7.3.1 Requirements

WAN is a means not only of data communications, but also of transport of interoffice voice and video traffic. First of all, we have to identify the business requirements.

Network engineers need to work with the client closely and find out both its business and its technical goals. For example, what kind of

Table 7.2 Sample Implementation Plan Worksheet

ID	ABC company systematic wan planning	Start	Finish	Duration
1	Business requirement analysis	04.04.2008	04.04.2008	11d
2	Meeting with head of it	15.04.2008	15.04.2008	1d
3	Meeting with financial manager	10.04.2008	10.04.2008	1d
4	Monitoring existing wan with network tools	10.04.2008	16.04.2008	7d
5	First assessment report	16.04.2008	16.04.2008	1d
6	Gathering the data	10.04.2008	19.04.2008	10d
7	Clarify the three key fundamentals	19.04.2008	23.04.2008	5d
8	Develop alternative topologies	20.04.2008	24.04.2008	5d
9	Selecting the best solution	20.04.2008	21.04.2008	2d
10	Final presentation	21.04.2008	21.04.2008	1d
11	Prepare the final project plan	21.04.2008	22.04.2008	2d
12	Start to implementation	21.04.2008	10.05.2008	20d
13	Job assignments	22.04.2008	22.04.2008	1d

The worksheet includes a Gantt chart for the year 2008 with a day-scale timeline (columns 4, 5, 6, 7, 8, 9, 10, 11, 12, 13, 14, 15, 16, 17, 18, 19, 20, 21, 22, 23).

new applications does the client want to add into the network because of new business requirements? How much network availability is required to support its core business operation? Accurately defining these goals is essential before starting the design work, because they are critical to the final success measurements of your job. It is beneficial for the later design phases if you initially understand the client's criteria for success and what goals must be met for the client to be satisfied.

For an enterprise WAN design, some of the typical business goals are to

- Increase company revenue and profit
- Increase employee productivity and improve corporate communication
- Reduce the telecommunication and network costs
- Improve the security of sensitive and proprietary corporate data
- Provide better customer support service
- Make data readily and securely available to all employees regardless of location
- Build partnerships with other companies

The most typical technical goals in an enterprise local area network (LAN)/WAN design include scalability, availability, performance, and security.

7.3.2 Constraints

Network engineers must carefully analyze business constraints such as

- Location constraints and geographical distribution
- Type of communications session/special applications or services in use (e.g., data, voice, video)
- Budget constraints

The combination of the first (requirements analysis) and second (constraints) fundamentals results in technology selection (or wide area network topology). Network engineers can decide and select available technologies, plan exactly where to place systems, and clarify the correct WAN topologies.

7.3.3 Presentation

The third fundamental of network planning is to write a proposal and communicate your ideas to others. Network engineers should develop a document that describes the business and technical requirements, the existing network, the logical and physical WAN design, and the budget and associated expenses. It should also include an executive summary and a primary project goal with all the details about the network topology, naming and addressing schemes, and security policies.

In general, the final design proposal should be comprehensive enough to cover the following topics:

- Executive summary
- Project goal
- Project scope
- Design requirements (both business and technical)
- Current state of the network
- Logical network design
- Physical network design
- Implementation plan
- Project budget

7.4 Five Sections

In the framework of the full planning pattern, for the second and third phases, you need to pass through the following five sections in order to develop the possible best WAN plan.

The short form condenses the four phases into six steps and combines Phase 2 and Phase 3. The short form is applicable to short or smaller planning assignments or situations.

7.4.1 Collect Key Data

Network engineers gather data on

- Business and technical needs
- Corporate structure
- Business information flow
- Applications in use/type of communication sessions

- Current topology
- Performance characteristics of current network
- Whether or not documented policies are in place
- Approved protocols and platforms

For example, the potential users of the WAN must be located and identified. They must be fairly accurately counted, and this number must be correlated to their physical location. The difficult part is estimating their propensity to consume bandwidth. Users will demand top-of-the-line everything in unlimited quantities—till they get the bill for it. Network planners, on the contrary, believe in an obscure law of physics that dictates all available bandwidth will immediately be consumed, regardless of the quantity supplied. One way of estimating the bandwidth requirements is to identify how the users are currently performing their work. If there are existing networks, such as X.25, asynchronous networks, or even modems, they can be invaluable sources of information. As Sportack et al. (1997) clarify, they should be monitored to determine

- Type of communications session (e.g., bulk data transfer, online transaction processing, Web access, videoconferencing, etc.)
- Frequency of use
- Peak use times
- Peak use traffic volumes
- Average duration of each session
- Average number of bytes transmitted per session
- Each user group's frequently accessed destinations

These are vital pieces of data that should form the core of your success, as the right WAN will be able to accommodate the projected traffic loads. In combination, these data reveal how much traffic will be put on the WAN and when it will be on the LAN. This is crucial in estimating the bandwidth required across every link of the network.

Another important piece of data is the required type of network performance. For example, will bulk data transfer constitute the majority of the traffic, or will interactive videoconferencing be the primary application? Is this situation likely to change in the near future? These two particular applications require opposite network performances. Bulk data transfer requires guaranteeing the integrity of the data delivered to its destination, regardless of the time it takes to get it there. Videoconferencing requires the network to deliver packets on time. Damaged packets are as worthless as late packets.

These details should be collected for each and every group of users that will be using the new WAN. Armed with this knowledge, the network planner can select the right WAN by considering the two primary aspects of wide area networking: technology and topology.

7.4.2 Investigate and Clarify Requirements

Getting data about requirements can be done with the help of Table 7.3. If a known/acceptable way is not available, seek help from others as to what data is wanted and how it can be obtained.

The relative importance of data, using the same AEIOU vowel-letter order-of-magnitude rating, is

A: Absolutely important
E: Especially important
I: Important issue
O: Ordinary important
U: Unimportant

7.4.3 Clarify Constraints and Relate to Requirements

The third section is clarifying constraints (locations, special applications or services, budget etc.) and relating to requirements. The output of this matching will come up with a WAN topology. We can use Table 7.4 to clarify this output (WAN topology). The Topology Worksheet consists of a Connections table.

The Connections table will clarify every location, number of users, special applications or services in use, adjacent network node, alternate route, capacity assumption, and available WAN technology. Locations are in every single point of branches. The number of users can be shown like this:

- 0–50 users⟶ L
- 50–100 users⟶ M
- Over 100 users⟶ H

Table 7.3 Requirements Analysis Worksheet

Project _____ No: _____

By _____ Date _____ Sheet of ―

What data to get		How to get	
Business requirements	Imp		Resp.
Consider and clarify project budget	A	Discuss with COO	
Reduce telecommunication and networkc osts	A	Discuss with COO	
Improve corporate communication	E	Discuss with COO	
Improve employee productivity	E	Discuss with COO	
Provide better customer support service		Discuss with COO	
Technical requirements	Imp		Resp.
Performance condition	A	Discuss with ITM	
Scalability concern	A	Discuss with ITM	
Reliability and security concerns	A	Discuss with ITM	
Implementation easiness	E	Discuss with ITM	
Convenience of maintenance and operational process	E	Discuss with ITM	

Reference notes: _____

We can indicate the type of communications session as follows:

- Data⟶ D
- Voice⟶ V
- Video⟶ E

Network engineers have to clarify point-to-point or point to multi-point connections for every location. The adjacent network node indicates the first network connection. Some locations need backup connections. The alternate route column shows the second network connection. The network engineer tries to define capacity assumptions based on the gathered data. As Wen (2011) summarizes, finally, the network engineer identifies WAN technologies as follows:

- Integrated services digital network (ISDN)
- Digital private lines

Table 7.4   Topology Worksheet

Project ————————————————————— No: —————————

By ————————————— Date ———————————— Sheet   of ——————

Connections

Locations	# of users	Type of comm.sess.	Adjacent network node	Alternate routes	Capacity assumption	Available WAN technology

Users

0 –50 users → L

50 –100 users → M

+100 users → H

Typeof Comm. Sess.

Data → D

Voice → V

Video → E

Reference notes: ——————————————————————————————

- Analog private lines
- Public packet networks
- Public frame networks
- Public automated teller machine (ATM) networks
- Broadband network technologies
- Digital subscriber line (DSL), asymmetric DSL (ADSL)
- Optical fiber
- Wireless, Institute of Electrical and Electronics Engineers (IEEE)802.11, local multipoint distribution service (LMDS), microwave
- Satellite technologies
- Network outsourcing

7.4.4 Integrate Presentation and Develop WAN Plans

The topology describes the way the transmission facilities are arranged. Wen (2011) points out numerous possible topologies (see Figures 7.2 through 7.11), and each one offers a slightly different mix of cost, performance, and scalability.

According to the gathered data, the network engineer tries to draw the network topology. The MS-Visio Tool (see Figure 7.12) can be used in this section, and several possible alternatives can be produced.

7.4.5 Evaluate and Accept the Best Plan

Selecting the *right* WAN is much more complicated than just picking technologies and a topology. Selecting the right WAN requires an understanding of the benefits and limitations of each topology and technology. This must be tempered with an assessment of each one's compatibility with other potential technologies. Other factors must also be considered during this process. The embedded base, budget constraints, skill sets, training costs, and even the scalability and

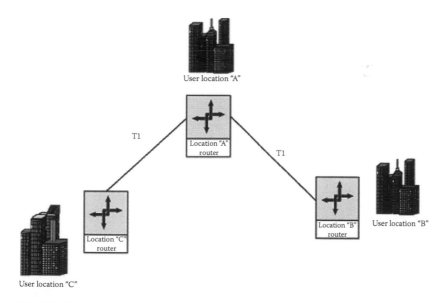

Figure 7.2 Peer-to-peer WAN topology. (Reprinted from Wen, Y., *Networking Enterprise IP LAN/ WAN Design*, The System Administration Company, San Jose, 2011.)

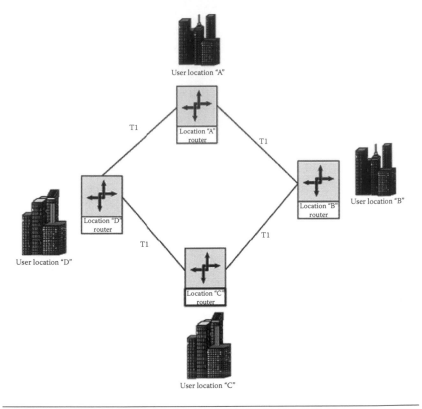

Figure 7.3 Ring WAN topology. (Reprinted from Wen, Y., *Networking Enterprise IP LAN/WAN Design*, The System Administration Company, San Jose, 2011.)

expected lifespan of each technology may all affect the selected WAN plan.

Each technology component must be carefully fitted to the network's topology. For example, using routing information protocol (RIP) on large, heavily trafficked, multitiered WANs would probably be a mismatch noticed by the user community.

Each decision that is made in the design phase has direct consequences for the functionality of the WAN. These consequences should be evaluated as carefully as the user requirements. For example, an important consideration is how much bandwidth each physical link in the WAN should provide. The consequences of this type of decision are easy to extrapolate. Transmission facilities incur monthly recurring charges that are mileage and/or bandwidth sensitive. Selecting too small a facility may save some money in the short run, but can cripple a company's ability to function.

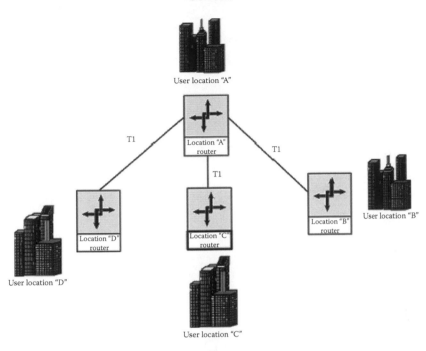

Figure 7.4 Star WAN topology. (Reprinted from Wen, Y., *Networking Enterprise IP LAN/WAN Design*, The System Administration Company, San Jose, 2011.)

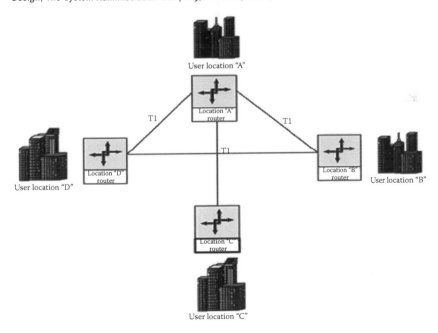

Figure 7.5 Partial mesh WAN topology. (Reprinted from Wen, Y., *Networking Enterprise IP LAN/ WAN Design*, The System Administration Company, San Jose, 2011.)

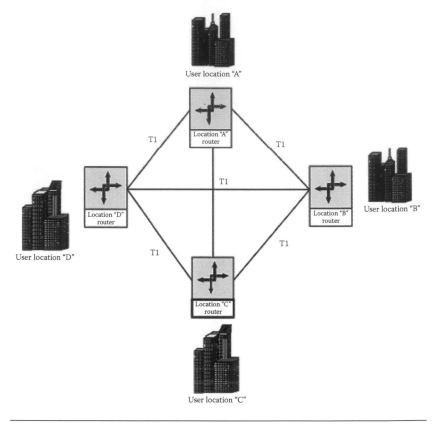

Figure 7.6   Full mesh WAN topology. (Reprinted from Wen, Y., *Networking Enterprise IP LAN/WAN Design*, The System Administration Company, San Jose, 2011.)

The last item to consider as you plan the WAN is the future. A well-designed WAN will not only satisfy its clients on its first day of operation; it will continue to satisfy them far into the future. This requires the network to be robust and flexible enough to accommodate technological changes, shifts in aggregate traffic patterns, and growth.

Remember, the WAN exists to facilitate the company's ability to conduct its business. Thus, its success should be measured more by the earnings potential it has created than by the costs it has incurred. With this in mind, study the technological and topological options. The right WAN is the one that delivers the performance your user base requires.

Use Table 7.5 for selecting and accepting the best plan. Identify each alternative appropriately, as described in Section 4.

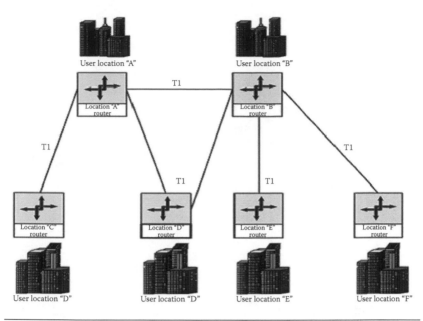

Figure 7.7 Top-tiered WAN topology. (Reprinted from Wen, Y., *Networking Enterprise IP LAN/WAN Design*, The System Administration Company, San Jose, 2011.)

Obtain, from suppliers or distributors of equipment and/or the company operations engineer, the cost of equipment and its installation. Determine the operating costs of each alternative plan. Prepare a worksheet, generated by Muther (2011), showing the comparable costs of each alternative. Also, on a separate similar worksheet, make a comparison of the intangible benefits and risks of each alternative. Compare alternatives, select the best, and get the others to approve it.

7.4.5.1 Comparison Based on Intangibles Enter the headings on a fresh copy of the worksheet, checking the box marked *intangibles* (upper left). Identify each alternative by a letter—X, Y, Z—and give a brief two-to-five-word description of each.

List all factors, considerations, or objectives the organization wants the project's intended plan to achieve. Select, or ask your approvers to select, the most important factors. Then, ask them to weigh the importance of each factor relative to the most important (10). Indicate each selected weight on the worksheet, and record by whom the weight values were determined.

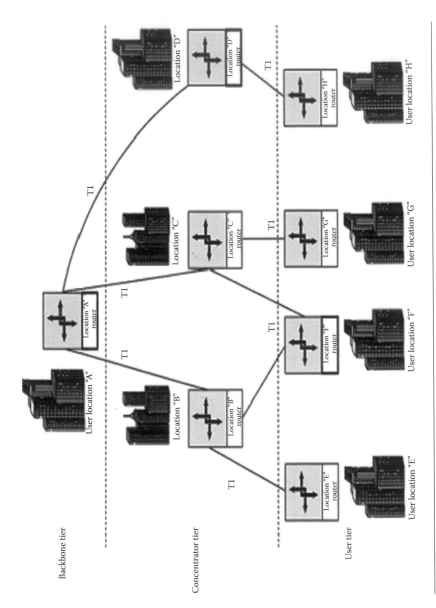

Figure 7.8 Three-tiered WAN topology. (Reprinted from Wen, Y., *Networking Enterprise IP LAN/WAN Design*, The System Administration Company, San Jose, 2011.)

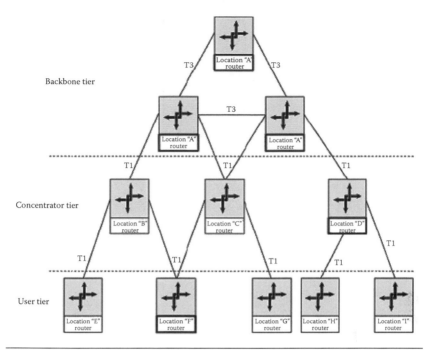

Figure 7.9 Multitiered hybrid WAN topology. (Reprinted from Wen, Y., *Networking Enterprise IP LAN/WAN Design*, The System Administration Company, San Jose, 2011.)

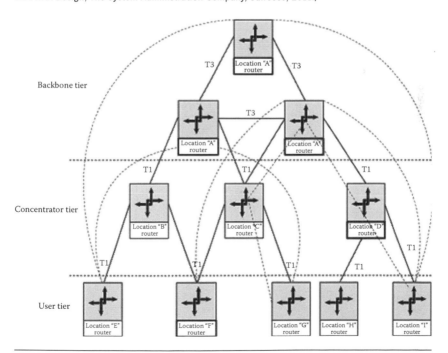

Figure 7.10 Three-tiered point-to-point WAN topology. (Reprinted from Wen, Y., *Networking Enterprise IP LAN/WAN Design*, The System Administration Company, San Jose, 2011.)

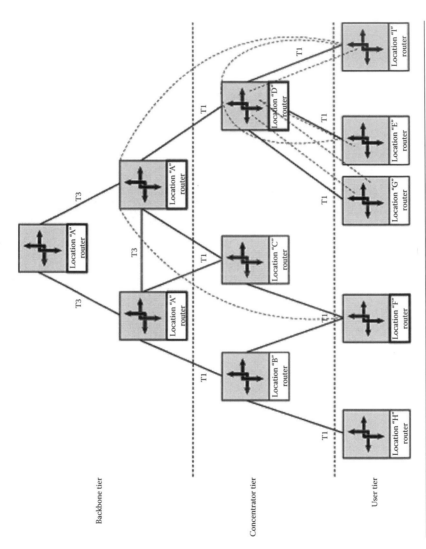

Figure 7.11 Traffic flow–based WAN topology. (Reprinted from Wen, Y., *Networking Enterprise IP LAN/WAN Design*, The System Administration Company, San Jose, 2011.)

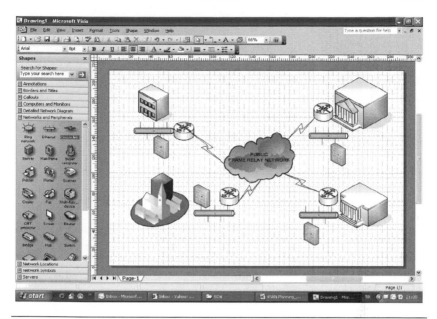

Figure 7.12 Network topology drawn by MS-Visio tool.

Ask your operations team and/or staff members who will use the proposed plan when installed to rate, for each factor, the effectiveness of each alternative in achieving that factor. Use A, E, I, O, or U to represent the descending order of effectiveness, as noted in the upper left-hand box of the worksheet. Enter, in the small rectangular *boxes within boxes* on the form, the selected vowel-letter ratings. Record the name(s) of the person(s) doing the rating.

After rating all alternatives for each factor, convert letters to numbers (A = 4, E = 3, I = 2, O = 1, U = 0) and multiply the rated number by the respective weight value. Enter the resulting weighted-rated values on the worksheet.

Down-total the weighted-rated values for each alternative, enter on the worksheet, and record by whom the tally was made. The alternative with the highest total should be the *winner*—subject to cost factors determined separately. In the lower left corner, indicate that these are weighted-rated down totals. Record any explanatory notations at the bottom, suitably referenced by an encircled lower-case letter. Typical intangible factors include

Table 7.5 Evaluation of Alternatives Worksheet

Evaluating alternatives		Project _____	Number _____

Costs:
Estimated by .... Approved by ....

Intangibles:
Weight set by .... Tally by ....
Ratings by .... Approved by ....

Evaluating description
A = Almost perfect, O = Ordinary pesult
E = Especially good, U = Unimportant results
I = Important result, X = Not acceptable

Project _____ Number _____
By _____ With _____
Date _____ Sheet _____ of _____

Description of alternatives:
X. _____
Y. _____
Z. _____
V. _____
W. _____

Factor/consideration	WT.	Alternative				
		X	Y	Z	V	W
1.						
2.						
3.						
4.						
5.						
6.						
7.						
8.						
9.						
10.						
11.						
12.						
13.						
14.						
15.						
Total	Annualized cost (line____ plus line____)					
	Weighted rated down total					

Reference notes:
a. _____
b. _____
c. _____
d. _____
e. _____
f. _____

Source: Muther, R., *Planning by Design*, Institute for High Performance Planners, Kansas City, 2011.

- Reduction in inventories and work in process
- Ability to respond quickly and with reliable service
- Reduction in operating effort
- Ease of effective supervision and/or worker convenience
- Use of machinery and production equipment

- Use of space
- Effectiveness of planning and control of work
- Effect on quality and avoidance of scrap/waste/rework
- Freedom from breakdown and maintenance attention
- Ease and speed of introduction of new methods or systems
- Freedom from disruption during installation
- Acceptance by key employees
- Freedom from personnel problems—availability of workers with proper skills, training capability, disposition of redundant workers, changes in job descriptions, union contacts, or work practices
- Enhancement of customer service ...

7.4.5.2 Comparison Based on Costs   In addition to the intangibles, the cost of investment and the operating costs for using and maintaining the project are also important. Identify each alternative by a letter—A, B, C—and give a brief three-to-five-word description of each.

List the names or titles of investment costs. For each alternative, down-total the investment costs, determine and enter the expected years of service life of the investment (the number of years the equipment is expected to operate), and divide the total investment cost by the years to determine the average annual investment cost. Typical investment costs include

- Equipment, new or rebuilt
- Transportation or travel costs
- Auxiliary equipment cost
- Area preparation
- Cost of moving and/or installation
- Planning and/or engineering services
- Training and run-in cost
- Freight in-bound for equipment
- Permits, excise tax cost

Then, list the titles for the operating costs. For each alternative, enter the estimated annual amount of each expense. Down-total this second group of costs. This gives the total annual operating cost for each alternative. Then, finally, add the average annual investment cost to the annual operating cost for each alternative. Typical operating costs include

- Direct material
- Scrap or waste
- Supplies and packing
- Maintenance or service contract
- Direct labor/salaries
- Fringe benefits
- Workers' compensation
- Insurance
- Power

This comparison will generally allow you to identify the lowest-cost alternative. However, it is not a financial justification. Should you need a more complete cost justification, ask your accountant for assistance.

8

SKYSCRAPER TERRORIST ATTACK RESPONSE PLANNING (STARP)

As networks increase in size and complexity, security issues and threats become more sophisticated and important. There is an increase in awareness of external threats, and consequently, organizations need better solutions to protect themselves from these threats. Security risk sources have changed a great deal due to the amazing progress in communication, international relations, information technology, and so on in recent years. They work in a more organized and planned fashion using these new methods, such that their sources and attempts are nearly impossible to be predict. In this respect, they become more capable of keeping their losses to a minimum.

This working model outlines a systematic methodology for planning responses to terrorist attacks on skyscrapers. Here, we provide a methodology for response planners to take the necessary precautions and apply them to their buildings. Skyscraper Terrorist Attack Response Planning (STARP) is a methodology for guiding and directing the management of emergency and disaster operations related to terrorist incidents and attacks. The methodology helps not only specific sections of the building, but also all levels and sections that need to be protected from terrorist attacks.

8.1 Introduction to STARP

Terrorist groups have changed their targets and methods from traditional patterns to methods that effect economic fluctuations, political instability, and cumulative civil losses.

Nowadays, the concept of security depends on the human factor and the implementation of security precautions. Furthermore, respect for human rights and ethics, customer satisfaction, and protecting the dignity of companies are also very important. Response planners need to take every important detail into consideration and take every necessary step.

The objective of this working model is to reduce the loss of human life and money due to terrorist attacks. Here, we provide a methodology for response planners to take the necessary precautions and apply them to their buildings. The planners assigned by top management, called *Company Security Officers* (CSOs), would be responsible for the application of this methodology.

STARP is a methodology for guiding and directing the management of emergency and disaster operations related to terrorist incidents and attacks. Figure 8.1 illustrates STARP—short version. The

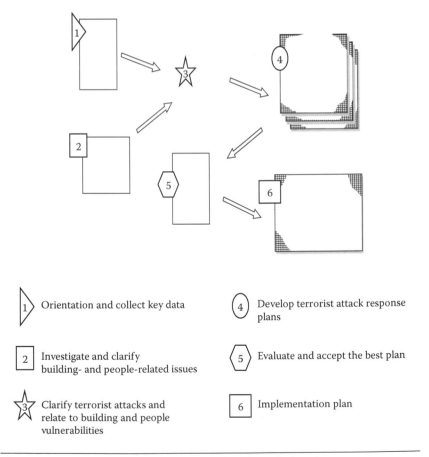

Figure 8.1   Skyscraper Terrorist Attack Response Planning—short version.

short version is applicable to short or smaller planning assignments or situations. The methodology helps not only specific sections of the building, but all levels and sections that need to be protected from terrorist attacks.

The Response Plan has been developed within the framework of the Standardized Emergency Management Procedures and has been integrated into several types of attacks and solutions.

8.2 Three Fundamentals

Three fundamentals of STARP are Building and People Issues, Probable Attacks, and Responses.

8.2.1 Building and People Issues

It is very important that you know what you are protecting. If you know your building well enough, then you are able to understand the weaknesses of the building and therefore, can take preventive actions against terrorist attacks.

People are more important than places, but bear in mind that a well-protected building can also protect human lives.

However, a well-protected building may mean nothing under a terrorist attack unless there are people who know what to do. In short, issues related with a well-protected building and a well-trained people should be known.

8.2.2 Probable Attacks

Terrorism is distinguished from other acts of violence and from war by always having these four characteristics:

- Terrorists violate the rules of modern warfare, established in acts called the Geneva Conventions and Hague Conventions, or they are actors who can't declare war legitimately.
- Their goal is to achieve political change.
- Their targets are symbols of the political issues in question.
- Acts of terrorism are designed to get attention from the public and the media.

The most important way to be protected from terrorist attacks is to have enough information about the probable threats.

8.2.3 Responses

Only a strong and well-protected building and well-trained personnel mean anything. Coordination between them is the key fundamental in solving the problem. You can only take the necessary precautions when you are aware of the possible threats; therefore, the second fundamental should be given serious and careful consideration. A response, called the *Terrorist Attack Response Plan*, should be prepared by integrating the first two fundamentals.

8.3 Six Steps

In the framework of the planning pattern, you need to pass through the following six steps to develop the best terrorist attack response plan.

8.3.1 Collect Key Data and Orientation

The main question of orientation is "What is our objective?" in this project. That is, we need to orient ourselves and to understand the project, the process, and the people involved. Then, we organize how we propose to plan and schedule the planning. The main issue is "What do we do?" and "How do we do it?"

- Understand the project: What? Why? Who? When? Where?
- Understand the purpose or objective(s), the external conditions, the situation(s), the scope/extent, any budget limitations, and the desired form of our planning output
- Understand and document the planning and people issues
- Make a schedule for the project planning

We can use the "Project Orientation Worksheet" designed by Muther (2011) (Table 8.1). This has three components:

- Project essentials
- Planning issues
- Planning schedule

Table 8.1 Project Orientation Worksheet

Description ———————————————— Project No: ————

Who is responsible? ———————— Authorized/Initiated by ——— Date ——

When project starts ———————— When planning starts ——— Sheet of —

Project essentials

1. Project objective(s) ——————————————————————
2. External condition(s) ——————————————————————
3. Situation(s) ——————————————————————
4. Scope/Extent ——————————————————————
5. Form of output ——————————————————————

Planning issues	Imp.	Resp.	Proposed resolution	Ok'd by
1.				
2.				
3.				
4.				

Planning schedule						Notes and Act
Task or action required to plan	Who					
1.						
2.						
3.						
4.						

Reference notes: ——————————————————————

Source: Muther, R., *Planning by Design*, Institute for High Performance Planners, Kansas City, 2011.

In "Project Essentials", enter

- The objective(s) or purpose(s) or goal(s) of this project
- The external conditions, such as synchronization with other projects, specific limitations, overall policies, or larger operational procedures …

- The situation(s): Physical, procedural, and personal situation circumstances
- The scope/extent of the project: How big? How detailed? When needed?
- The form of this planning's output: Written report? Action plan approved?

In "Planning Issues", we enter each problem, uncertainty, and question—one line for each—on the left. In the first column, record how important the issue is to this project. Here, we enter a vowel-letter as our order-of-magnitude judgmental rating:

A: Absolutely important
E: Especially important
I: Important issue
O: Ordinary important
U: Unimportant

In the "Responsible" column, enter who is responsible for getting the issue resolved, and mark the initials of the approver.

In "Planning Schedule," list each action required to plan what we intend to do to prepare a terrorist attack response plan. List one action on each line, and show who is responsible for doing it. Set a calendar schedule at the top of the vertical lines.

8.3.2 Investigate and Clarify Building- and People-Related Issues*

Our first fundamental is to investigate the issues about the building and people. Here, we have to investigate them separately. The response planners should take the issues related to the building and people in the following subsections as a guide for their assessment. But, planners should keep in mind that building- and people-related issues should not be limited to these items only.

---

* Adapted from Atlas (2013).

8.3.2.1 Building Related

1. *Identify building entrances:* One of the vulnerabilities of a building is its entrances. Well-protected entrances may dissuade the terrorist from attack. First, the user must identify all the possible entrances, such as doors, windows, roof, garage, other openings, ventilation ducts, and so on.

2. *Identify main entrances:* "Main entrance" means one or more entrances that are used by everyone to enter the building. Generally, there is one main entrance for a building. Sometimes, entry is possible from the garage. If so, the garage is also to be counted as a main entrance. Everyone must go through a security check when entering the building. Install X-ray machines.

3. *Mark and identify emergency escape routes:* To evacuate the building in the case of a terrorist attack, identifying emergency escape routes is very important. Draw emergency escape routes on building plans. Emergency escape routes and exit doors that are not in common use should be clearly indicated, as appropriate, by suitable signs. However, in certain circumstances, such as places of public assembly, you should indicate all exit doors. All signs should be in positions where they can be seen clearly. These signs must take the form of a pictogram, which may incorporate a directional arrow. The sign can also be supplemented by words such as "Fire Exit."

    Note: Fire safety signs must comply with the relevant requirements of the Health and Safety (Safety Signs and Signals) Regulations 1996.

4. *Plan evacuation procedures:* Write an evacuation plan procedure and distribute it to every employee in the building.

5. *Ensure complete evacuation:* Be sure that everyone in the building has been evacuated. Prepare checklists for group leaders.

6. *Identify outside meeting places:* After evacuation, there should be marked places where people will gather and take the necessary precautions.

7. *Method of transporting injured or incapacitated persons:* During evacuation, there may be injured people. The method should be carefully selected.

8. *Method of providing care/implementing emergency procedures if provider is injured or incapacitated:* Everyone should have a substitute, especially first aiders, because they are the key personnel.

9. *Identify what kinds of things are allowed to enter the building:* Identify what kinds of things could be brought into the building (such as personal guns, deodorants, powders, and so on) and notify the relevant parties.

10. *Identify the doors and elevators to be used by access cards:* Some doors and elevators can be set aside for limited access. Identify these doors and elevators clearly and mark them as being for authorized personnel access only.

11. *Identify garage entrances and precautions to be taken:* Garage entrances are one of the weakest points. Extra care should be taken, and a security procedure should be prepared.

12. *Decide where to put the cameras (outdoor and indoor):* Get professional help for this. There should be no blind spots. Cameras should be placed carefully.

13. *Identify restricted areas: Restricted area* means an area to which access is limited by the licensee or registrant for the purposes of protection of individuals against undue risks from exposure to radiation and radioactive materials. A restricted area shall not include areas used as residential quarters, although a separate room or rooms in a residential building may be set apart as a restricted area.

14. *Identify personnel access card levels:* All employee and visitor access cards should have levels, and entrance to restricted areas should be restricted by access card levels, such as Access Level 1–3.

15. *Establish an access card system:* Using this system, the user may monitor the entrance and exit times of everyone in the building.

16. *Establish a method for alerting people and local authorities:* Preferably, this will be a loud alarm system such as a general alarm system or a fire alarm system. But for local authorities, there may be hidden alarm systems at certain points. Decide where to place them.

17. *Choose which fire detector to use:* Again, professional help is strongly recommended here. Choose whether smoke or heat detectors should be used.

18. *Determine firefighting equipment type:* Again, professional help is strongly recommended here. There are many types, usually sold as a set. Decide which set to supply.

19. *Select where to install detectors and sprinkler systems:* A sprinkler system should be established for extinguishing fires. But, the most important thing here is choice of location. A detector can detect in an area up to 30 m².

20. *Identify the supply entrances:* Another weak point, like the garage, is supply entrances. There should be extra checks for supplies from outside.

21. *Determine where to place the Emergency Escape Breathing Devices (EEBDs):* An EEBD is a supplied air or oxygen device only used for escape from a compartment that has a hazardous atmosphere, and should be of an approved type. Place them beside the entrances on each floor.

22. *Install movement sensors:* The user may switch on movement sensors after hours in some key places.

23. *Identify the location of the jammer:* This is not needed by all buildings, but Level 5 buildings will need a jammer. A jammer is an active electronic countermeasures (ECM) device designed to deny intelligence to unfriendly detectors or to disrupt communications.

24. *Erect barriers to building entrances:* Barriers will not allow cars to get close to the building; therefore, the impact area of a car bombing is automatically reduced. Decide what kind of barrier to use.

25. *Determine the risky companies in the building:* This is another interesting point. The target may not be your building but a company in the building. As there are many companies in a skyscraper, the other companies will be affected. As an example, no one attacks a shoeshine saloon, but they may attack a bank branch nearby, which will affect them both.

26. *Identify the materials to be rescued in case of emergency:* There may be very valuable records and documents in the building. Identify them clearly and choose a method of marking them.

27. *Define items that should not be on an escape route:* You should make sure that items that pose a potential fire hazard or those that could cause an obstruction are not located in corridors or stairways intended for use as a means of escape.
28. *Identify ways of communication:* Walkie-talkies, and so on.

8.3.2.2 People Related

1. *Determine the security company:* Always work with professionals. Do not select a security firm on the basis of price alone. Undertake the necessary investigations. Ask questions. Select a company that is certified and well recommended.
2. *Develop emergency response plans:* Develop emergency response plans and hang them on every floor where people may reach them at any time. Decide the method.
3. *Prepare "muster cards" for personnel:* Prepare muster cards for all personnel showing their duties in the case of emergency situations. Ensure that personnel always keep them.
4. *Prepare emergency situation job descriptions:* Muster cards are brief job descriptions. Detailed job descriptions should be prepared and distributed to employees.
5. *Determine substitutes for key personnel:* In emergency situations, all key personnel should have substitutes. The master plan should continue perfectly in emergency areas.
6. *Determine team leaders:* A team leader motivates the people in the team, directs them, and makes important decisions that may change the destiny of the group.
7. *Conduct expected and unexpected drills:* Training is crucial for continual improvement. Training should be done at least yearly, and unexpected drills should also be conducted. Unexpected drills measure the reactions of people in emergency conditions and reflect a more realistic situation.
8. *Introduce guard rounds:* During the night, or at certain times depending on the type of building and threat, make the guards check around for any abnormal situations.
9. *Determine the number of employees:* This part is also so important to arrange teams and some other related items written above is so much dependent on the number of employees.

10. *Determine the number of companies:* As for Item 9, and in addition, this is important for sharing the security expenses.

11. *Determine the number of daily visitors:* As for Item 9. The entrance of the building is designed according to the number of visitors.

12. *Determine the numbers to be called in case of emergency:* Every person in the building, especially guards and key personnel, should be able to reach a list including numbers to be called in case of emergency.

8.3.3 Clarify Terrorist Attacks and Relate to Building and People Vulnerabilities

Terrorism is a controversial term with no internationally agreed single definition. In the modern sense, it is violence against civilians to achieve political or ideological objectives by creating fear. Most common definitions of terrorism include only those acts that are intended to create fear (terror), are perpetrated for an ideological goal (as opposed to a lone attack), and deliberately target or disregard the safety of noncombatants. Some definitions also include acts of unlawful violence and war.

Terrorism is also a form of unconventional and psychological warfare. The word is politically and emotionally charged, and this greatly compounds the difficulty of providing a precise definition. A person who practices terrorism is a terrorist.

Terrorism has been used by a broad array of political organizations in furthering their objectives: both right-wing and left-wing political parties, nationalistic and religious groups, revolutionaries, and ruling governments. The presence of non-state actors in widespread armed conflict has created controversy regarding the application of the laws of war.

An International Round Table on Constructing Peace, Deconstructing Terror (2004) hosted by the Strategic Foresight Group recommended that a distinction should be made between terrorism and acts of terror. While acts of terrorism are criminal acts as per the United Nations Security Council Resolution 1373 and the domestic jurisprudence of almost all countries in the world, terrorism refers to a phenomenon including the actual acts, the perpetrators of acts of terrorism themselves, and their motives. There is disagreement on definitions of terrorism. However, there is an intellectual consensus globally that acts of terrorism should not be accepted under any circumstances. This is reflected in all important conventions, including the United Nations counterterrorism strategy, the decisions of the Madrid Conference on terrorism, the Strategic Foresight Group, and the Alliance of Liberals and Democrats for Europe (ALDE) Round Tables at the European Parliament.

In this step, planners should use the *Vulnerability Assessment Form* (see Table 8.2) to point out the vulnerabilities of the building and the possible ThreatCon. With the help of this form, planners can determine the expected/possible kind of terrorist attack. Planners should also determine in which areas of the building this kind of attack is expected, and accordingly, its impact should be entered. Plus, the existing condition of the building should also be marked. After that, planners should fill in the importance and existing frequency and then multiply them and find out the Total Risk Point (TRP). Consequently, each possible terrorist attack TRP is added, and the TRP of the building is found, which helps to determine the ThreatCon of the building in the following step.

Some terrorist attack examples are listed in the following subsection. Planners can take these issues as a guide for their study, but they should also keep in mind that terrorist attacks are not limited to these items.

*8.3.3.1 Terrorist Groups by Type**   Terrorism is best understood as a modern phenomenon: as violent struggle between non-state organizations and modern states, and because it relies on mass media to spread terror among as many people as possible. However, there are some premier groups who use terror to achieve political ends, and who are often considered precursors to modern terrorists:

1. *Socialist/communist:* Many groups committed to socialist revolution or the establishment of socialist or communist states arose in the second half of the twentieth century, and many are now defunct. The most prominent are

---

* Adapted from Mahan and Griset (2013).

Table 8.2 Vulnerability Assessment Worksheet—Example

Vulnerability assessment form

Description TO KNOW THE VULNERABILITIES IS VERY IMPORTANT FOR SOMEONE TO TAKE PREVENTIVE ACTIONS_____   Project No 001/08 _____

Who is responsible? CSO_____   Authorized/Initiated by GM_____   Date 03RD JUNE 2008 _____

When is project due? 07TH JUNE 2008_____   When is Planning Due 14TH JUNE 2008_____   Sheet 3 _____ of 6 _____

General vulnerability assessment

No	Expected terrorist attack	Expected area	Dimension	Impact	Existing Condition			Importance	Existing frequency	Risk point
					Normal	Abnormal	Urgent			
1	Chemical biological terrorist attacks	Inside of building	Chemical and biological residues	Loss of lives no damage to building		.	.	5	1	5
2	Nuclear terrorist attacks	Outside or inside of building	Nuclear residues	Loss of lives and damage to building		.	.	5	1	5
3	Car bombing	Outside of building	.	Loss of lives and damage to building	.		.	3	2	6
4	Dirty bomb	Outside or inside of building	Nuclear residues	Loss of lives and damage to building		.	.	4	1	4
5	Assassination	Outside or inside of building	.	Panic in the public		.	.	3	1	3
6	Rocket propelled grenades	Outside of building	.	Loss of lives and damage to building		.	.	2	1	2
7	Improvised explosive devices	Outside or inside of building	.	Loss of lives and damage to building		.	.	3	1	3
8	Cyber terrorism	Outside of building	Loss of reputation	Damage to it systems loss of time and money	.		.	2	2	4
9	Suicide bomber	Outside or inside of building	.	Loss of lives and damage to building	.		.	4	2	8
										0
										0
										0
										0
										0
										0
Guide									Total risk point	40

Importance	Points		Existing frequency (In Area)	Points		Total risk point (Imp X Freq)
A - Absolutely important	5 PTS		Very frequent	5 PTS		Very risky = Threatcon delta
E - Especially important	4 PTS		Frequent	4 PTS		Risky = Threatcon charlie
I - Important issue	3 PTS		Not so frequent	3 PTS		Medium risk = Threatcon beta
O - Ordinary importance	2 PTS		Rare	2 PTS		Unrisky = Threatcon alpha
U – Unimportant	1 PTS		Very rare	1 PTS		Normal = Threatcon normal
Reference notes:						

- Baader-Meinhof Group (renamed Red Army Faction, defunct as of 1998) (Germany)
- Popular Front for the Liberation of Palestine (PFLP)
- Red Brigades (Italy)
- Revolutionary Struggle (Greece)
- Shining Path (Peru)
- Weather Underground Organization (United States)

2. *National liberation:* National liberation is historically among the most potent reasons why extremist groups turn to violence to achieve their aims. There are many of these groups, such as
   - ETA (Basque)
   - Fatah (PLO) (Palestinian)
   - Irgun (Zionist)
   - IRA (Irish)
   - PKK (Kurdish)

3. *Religious-political:* There has been a rise in religiosity globally since the 1970s and with it, a rise in what many analysts call religious terrorism. It would be more accurate to call groups such as Al Qaeda religious-political or religious-nationalist. We call them religious because they use a religious idiom and shape their "mandate" in divine terms. Their goals, however, are political recognition, power, territory, concessions from states, and the like. Such groups are
   - Al Qaeda (transnational, Islamist)
   - Aum Shinrikyo (renamed Aleph) (Japanese, various influences including Hindu and Buddhist)
   - Ku Klux Klan (U.S., Christian)
   - Abu Sayyaf (Philippines, Islamist)
   - Egyptian Islamic Jihad
   - Hamas (Hamas is designated by the U.S. and other governments as a terrorist group, but it is also the elected government of the Palestinian Authority)
   - Hezbollah (it is designated as a terrorist organization by the U.S. and other governments, but others argue it should be considered as a movement)

4. *State terrorism:* Most states and transnational organizations (such as the United Nations) define terrorists as non-state actors. This is often a highly contentious issue, and there are

long-standing debates in the international sphere over a few states in particular, including the United States. There are some states or state actions in history over which there is no dispute, though, such as Nazi Germany or Stalinist Russia.

8.3.3.2 Kinds of Terrorist Attacks*

1. *Chemical-biological terrorist attacks:* The types of terrorist attacks are limited only by the terrorists' imagination and by the safeguards we throw up to block them. Analysts are always saying how *adaptive* the terrorists are. It's true. Small, compartmentalized groups fueled by a common hatred and ideology have all the time in the world to hatch conspiracies and to act on them.

2. *Nuclear terrorist attacks:* A nuclear attack is the type of terrorist attack most feared by the public. The horror and damage of a nuclear explosion on American soil would therefore gain the terrorists the most publicity and, yes, terror. Terrorism analysts say that the complexity of nuclear delivery systems has so far kept that type of attack less likely than a chemical-biological attack or a *dirty bomb*.

3. *Car bombing:* Car bombs, which are called Vehicle-Borne Improvised Explosive Devices by the military, use explosives to weaponize cars, trucks, and even motorcycles. They are used globally by terrorists and militias in assassinations aimed at killing a specific individual and in attacks designed to achieve mass destruction of people and property.

   In some car bombings, explosives are rigged to a car or truck's ignition system, and triggered when the vehicle is turned on. In others, explosives are attached to another part of the car, or beneath it, and set off remotely.

   Car bombs can also be deployed in suicide attacks. In such attacks, a car or truck is packed with explosives and then driven into a building or another vehicle, and so on.

4. *Dirty bomb:* Dirty bomb is the colloquial name given to a Radiological Dispersal Device, a bomb that combines a conventional explosive with radioactive material. The conventional

---

* Adapted from Johnson (2013).

explosive embeds radioactive material within it, so that when it is detonated, it disperses the radioactive agent. The U.S. Nuclear Regulatory Commission has indicated that most dirty bombs would not contain enough radiation to cause significant lethal damage. Thus, although they use radioactive materials, dirty bombs are "in no way similar to a nuclear weapon." Their primary use is to terrorize, rather than cause widespread damage.

Another way of dispersing radioactive materials is to simply leave them somewhere where they will be released into the air, "creating fear and possibly panic, and requiring potentially costly clean-up."

5. *Assassination:* Assassination is the term given to the murder of political or other well-known figures. It has been a stock ingredient in terrorists' arsenal since *Narodnaya Volya* (The People's Will), a late eighteenth-century Russian group, assassinated Tsar Alexander II, a symbol of the feudal system they wished to revolutionize.

Notable assassinations by terrorist groups include the 1981 assassination of Egyptian President Anwar Sadat by Egyptian Islamic Jihad to protest against Sadat's normalization of relations with Israel. In 1995, Israeli Prime Minister Yitzhak Rabin was assassinated by Yigal Amir, an extremist orthodox Jew who believed there was divine justification for killing a head of state.

6. *Rocket-propelled grenades:* The lightweight, shoulder-launched weapons known as RPGs were originally designed to damage tanks. Their name, in fact, means *handheld anti-tank grenade launcher* in Russian, the language of its original Soviet manufacturers (*Ruchnoy Protivotankovy Granatomyot*). In common English usage, RPG is understood to stand for *rocket-propelled grenade*. Those most widely in use today are RPG-7s, first developed by the Soviets.

Despite their originally intended use as anti-tank weapons, RPGs inflict most damage on unarmed vehicles and personnel in conflicts today, since tanks have been designed to resist RPG attacks. RPGs' easy portability, low cost, and wide availability in the black markets of the Middle East and Eastern Europe make them popular among terrorist groups and other substate militias.

7. *IEDs (Improvised Explosive Devices):* The main distinguishing feature of an improvised explosive device, commonly called an IED, is its makeshift or homemade quality. IED builders use materials at hand, and a variety of techniques, to create explosives designed to be lethal. They may use explosives alone or combine explosives with radiological, chemical, or biological materials. The explosives may be homemade, commercial, or military grade. IEDs are composed of
   - An initiation system that sets off the explosive
   - Explosive material
   - A detonator
   - A container or means of conveyance

   IEDs may be detonated in one of several ways, depending on their design and intended target. Those that use concealed mortar and artillery projectiles can be thrown into or placed in a designated area, often hidden.

   Vehicle-borne IEDs (VBIEDs) use cars or trucks to contain the explosive device. Suicide bomb IEDs make use of the human body to convey the IED. The Vietcong made use of IEDs during the Vietnam War. Other militias and terrorist groups, including Chechen rebels, Hezbollah, the IRA, and Iraqi insurgents, have used IEDs in attacks. In October 2006, the U.S. military reported that the number of IEDs in Iraq was higher than at any previous point.

8. *AK-47 assault rifles:* The AK-47 assault rifle was developed by Russian national Mikhail Kalashnikov in 1947. Kalashnikov was a soldier in World War II and as a result of his first-hand experience in battle with the Nazis, who were well equipped with submachine guns, he determined to create a firearm that would guarantee Soviet superiority in war.

   By refining and combining elements of existing automatic weapons and assault rifles, Kalashnikov created a weapon that was simply designed, lightweight, and easy for even poorly trained soldiers to aim, without sacrificing any of a machine gun's lethal power. These qualities have made the AK-47 an effective weapon for use by paramilitaries and in urban warfare.

   The Soviet Army adopted AK-47s in 1949. During the Cold War, AK-47s (and the next generation of Kalashnikovs,

AKMs) were used by communist armies throughout the world. The Soviets also distributed them to leftist guerrilla armies or militant groups who served or supported Soviet interests.

According to the United Nations Office on Drugs and Crime, "30–50 million copies and variations of the AK-47 have been produced globally, making it the most widely used rifle in the world." They continue to be sought after by terrorist groups and paramilitaries, as well as gangs and drug dealers.

9. *Cyber-terrorism:* Many businesses and government activities have come to depend on information systems. It is also expected that there will be an acceleration in the use of information technologies and networking. For critical infrastructure, such as power supply, transportation, and electronic control, information systems also have a crucial role in maintaining public safety and stable supplies of indispensable services for the economic activities of business and the daily lives of the people.

An electronic attack using telecommunications networks and information systems (called a *cyber-attack*) on these important information systems, which are fundamental to the critical infrastructure, has the potential to disrupt people's lives and business activities, as well as to cause large amounts of damage, placing people's lives at risk. This kind of attack, unlike a physical attack, can be made from a single computer by a person with the ability to intrude into the information system. There is also the fear that systematic, large-scale attacks could be made for the purpose of causing disruption and confusion to business activities and people's lives.

Overseas, there have been cases of damage to financial information systems, individuals known as hackers intruding into critical information systems, and denial-of-service (DoS) attacks, as well as instances of large amounts of damage caused by the spread of computer viruses; so it is clear that this threat is already becoming a reality. The United States is developing a congressional plan to handle the threat of financial damage, confusion, injury, or death caused by attacks on critical networks by terrorists/criminal organizations with advanced technical skills.

Connections via the internet and other networks continue to develop, and interdependence is increasing. There is also increasing standardization and commonality in the specifications of information systems. These trends increase the threat of cyber-attacks, even on information systems that currently face little danger from outside intrusion. In addition, there is always the possibility of such attacks being made by inside personnel. It must be recognized that even an information system that is not connected to any other networks is not immune to the danger of an outside attack.

8.3.4 Develop Terrorist Attack Response Plans

After determining the riskiness of the building in the previous step, the planner should enter the terrorism threat conditions (ThreatCon). The planner should also select the level of the building from Level 1 to Level 5, as described in the following subsection.

Accordingly, selecting the above conditions, the planner is ready to use the *Terrorist Attack Response Plan* (see Table 8.3), which aims to find out what precautions need to be taken to prevent or minimize losses.

8.3.4.1 Building Levels

- *Level 1:* A Level 1 facility has 50 or fewer employees. In addition, the facility is likely less than 5000 m² or less than 10 floors.
- *Level 2:* A Level 2 facility has between 50 and 100 employees. In addition, the facility is likely between 5000 and 7500 m² or between 10 and 15 floors.
- *Level 3:* A Level 3 facility has between 100 and 250 employees. In addition, the facility is likely between 7500 and 10,000 m² or between 15 and 20 floors. It has a moderate to high volume of public contact.
- *Level 4:* A Level 4 facility has between 250 and 500 employees. In addition, the facility is likely between 10,000 and 12,500 m² or between 20 and 25 floors. Has a high volume of public contact.
- *Level 5:* A Level 5 facility has more than 500 employees. In addition, the facility is likely more than 12,500 m² or more than 25 floors. Has a high volume of public contact.

Table 8.3　Terrorist Attack Response Plan Worksheet—Example

Terrorist attack response plan

Status reported as of 010th June 2008

Expected terrorist attacks

Total vulnerability point

Project skyscrapon terrorist attack response plan

By company security officer

Date 010RD June 2008

Number 001/08

With

Sheet 4　　　of 6

Building level 1 | Building level 2 | Building level 3 | Building level 4 | Building level 5

ThreatCon normal | ThreatCon alpha | ThreatCon beta | ThreatCon charlie | ThreatCon delta

Preliminary screening for alternatives

1　Day / Night camera

Key to threatcons
Very risky – Threatcon delta　80×TRP×125
Risky – Threatcon charlie　40×TRP×80
Medium risk – Threatcon beta　20×TRP×45
Unrisky – Threatcon alpha　5×TRP×20
Normal – Threatcon normal　0×TRP×5

Key to building level
Level I　A level I facility has 50 or fewer employees. In addition, the facility likely has less than 5000 sq metres or less than 10 floors
Level II　A level II facility has between 50 - 100 employees. In addition, the facility likely has between 5000 sq metres - 7500 sq metres or between 10 - 15 floors
Level III　A level III facility has between 100 - 200 employees. In addition, the facility likely has between 7500 sq metres - 10000 sq metres or between 15 - 20 floors. Has a moderate to high volume of public contact.
Level IV　A level IV facility has between 200 - 500 employees. In addition, the facility likely has between 10000 sq metres - 12500 sq metres or between 20 - 25 floors; high volume of public contact.
Level V　A level V facility has more than 500 employees. In addition, the facility likely has more than 12500 sq metres or more than 25 floors; high volume of public contact.

Keys to evaluation
A Absolutely important
E Especially important
I Important issue
O Ordinary important
U Unimportant

Reference note:

a.

b.

c.

d.

8.3.4.2 Terrorism Threat Conditions (THREATCON) When a threat of possible terrorist activity exists, the following preventative measures and actions should be taken by the workplace supervisors and individual employees (Johnson 2013):

- *THREATCON NORMAL:* When a threat of possible terrorist activity exists but only warrants a routine security posture.
- *THREATCON ALPHA:* Threat of possible terrorist activity against personnel and facilities, the nature and extent of which are unpredictable, and circumstances do not justify full implementation of the next level (Bravo).
- *THREATCON BRAVO:* There exists an increased and more predictable threat of terrorist activity.
- *THREATCON CHARLIE:* An incident occurs, or intelligence is received indicating that some form of terrorist action against personnel and facilities is imminent.
- *THREATCON DELTA:* This is normally a localized condition when a terrorist attack has occurred or when intelligence has been received that terrorist action against a specific location or person is likely.

8.3.5 Evaluate and Accept the Best Plan

Preliminary screening for alternatives found in the Terrorist Attack Response Plan should be separately evaluated, and the best applicable one should be selected using the Evaluation of Alternatives Worksheet.

Identify each alternative appropriately, as described in Step 4. Prepare a worksheet showing the comparable costs of each alternative. Also, on a separate worksheet, make a comparison of the intangible benefits and risks of each alternative. Compare alternatives, select the best, and get the others to approve it.

Enter the headings on a fresh copy of the worksheet (see Table 8.4) generated by Muther (2011), checking the box marked *intangibles* (upper left). Identify each alternative by a letter—X, Y, Z—and give a brief two-to-five-word description of each.

List all factors, considerations, or objectives the organization wants the project's intended plan to achieve. Select, or ask your approvers

Table 8.4   Evaluation of Alternatives Worksheet

Evaluating alternatives					

☐ Costs
  Estimated by _____ Approved by _____
Total Annualized Cost = Investment Cost ____ +Annual Operating Cost
                        Expected Life

☐ Intangibles
  Weight set by _____ Tally by _____
  Ratings by _____ Approved by _____
            Evaluating description

A	Almost perfect	O	Ordinary results
E	Especially good	U	Unimportant results
I	Important results	X	Not acceptable

Project _____   Number _____
By _____   With _____
Date   Sheet _____

Description of alternatives

X. _____
Y. _____
Z. _____
V. _____
W. _____

Factor / consideration	WT	Alternative				
		X	Y	Z	V	W
1						
2						
3						
4						
5						
6						
7						
8						
9						
10						
11						
12						
13						
14						
15						

Total ☐ Annualized cost (line ____ Plus line ____)
       ☐ Weighted rated down total

Reference notes:
a. _____   d. _____
b. _____   e. _____
c. _____   f. _____

Source: Muther, R., *Planning by Design*, Institute for High Performance Planners, Kansas City, 2011.

to select, the most important factors. Then, ask them to weigh the importance of each factor relative to the most important (10). Indicate each selected weight on the worksheet, and record by whom the weight values were determined.

Ask your operations team and/or staff members who will use the proposed plan when installed to rate, for each factor, the effectiveness of each alternative in achieving that factor. Use A, E, I, O, or U to represent the descending order of effectiveness, as noted in the upper left-hand box of the worksheet. Enter, in the small rectangular *boxes within boxes* on the form, the selected vowel-letter ratings. Record the name(s) of the person(s) doing the rating.

After rating all alternatives for each factor, convert letters to numbers (A = 4, E = 3, I = 2, O = 1, U = 0) and multiply the rated number by the respective weight value. Enter the resulting weighted-rated values on the worksheet.

Down-total the weighted-rated values for each alternative, enter on the worksheet, and record by whom the tally was made. The alternative with the highest total should be the *winner*—subject to cost factors determined separately. In the lower left corner, indicate that these are weighted-rated down totals. Record any explanatory notations at the bottom, suitably referenced by an encircled lower-case letter.

8.3.6 Implementation Plan

This step is dedicated to carrying out the selected plan. In implementation, who does what is very important. It includes the actions needed to make the plan come true. Who will be responsible and the duration of action have to be clarified. This step sets the framework for dealing with the *expectations* on time. We can use the MS Project tool in this step (see Table 8.5).

Table 8.5   Implementation Plan Worksheet

Project plan and schedule Status reported as of _____	Project _____ By _____ Date _____	Number _____ With _____ Sheet _____ of _____

Work to do; action to take	Resp. of	Calendar time	Further schedule
1			
2			
3			
4			
5			
6			
7			
8			
9			
10			
11			
12			
13			
14			
15			
16			
17			
18			
19			
20			
21			
22			
22			
24			
25			

Gantt chart code: Each vertical line is one unit of time

Reference notes:

☐ Earliest date work scheduled to start   a. _____

☐ Latest date work scheduled to finish   b. _____

☐ Duration of scheduled work/task   c. _____

■ Percent completion as heavy filled-in line   d. _____

Source: Muther, R., *Planning by Design*, Institute for High Performance Planners, Kansas City, 2011.

9

SYSTEMATIC PLANNING OF GLOBALIZING LOCAL FIRMS (SPG)

HAKAN BÜTÜNER AND ALI AMJAD QAZI

This working model outlines a systematic planning methodology for helping local businesses, which in some way have the potential to grow in the international market but feel insecure regarding the risks and threats involved, to grow globally, and will further create a sense of motivation and ambition. In this chapter, we try to solve their issues using a systematic procedure by considering three basic fundamentals to further enhance their horizons with the relevant knowledge. Figure 9.1 illustrates SPG-short version.

9.1 Introduction to SPG

Over the last 20 years, the world has become a much smaller place, as people around the world are more connected to one another through the rapid flow of information and money and as goods and services produced in one part of the world are easily accessible in other parts of the world. In simple terms, it is the economic and cultural communication between various countries around the world. According to the World Bank, globalization is the growing integration of economies and societies around the world. It is the process of interaction and integration among people, companies, and governments from different countries through international trade and investment. It allows a country to perform international trade and business and to enjoy the expansion of market and trade links between various countries around the globe.

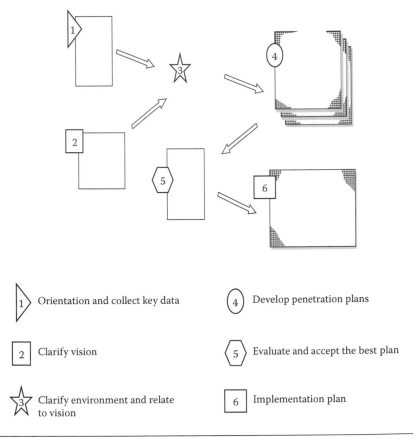

Figure 9.1   Systematic planning of globalizing local firms.

Our main objective here is to boost the confidence of local companies by introducing a systematic planning tool that they can use which helps them to give the chance to become global and compete in the international market.

9.2 Three Fundamentals

The three fundamentals of SPG are Vision, Environment, and Penetration.

9.2.1 Vision

Every company should have a clearly defined vision and mission. Once the frame of its vision is well understood, then basic research is conducted to identify the firm's potential target.

9.2.2 Environment

A business tends to explore its further potential while competing with its local competitors by using all kind of marketing activities and focusing on managerial issues to grow further in the market. After becoming successful in any particular industry locally, companies want to grow in the international market. Normally, however, companies hesitate due to the language barriers, foreign business policies, taxation, currency fluctuations, and interest rates.

Accordingly, both the macro and micro environments of the particular country need to be taken into consideration by every company. According to Porter (1998), macro-environmental factors are political, economic, social, technological, and legal. It is vital to analyze macro-environmental variables that are likely to influence the business's demand and supply levels and costs. Besides, it is also important to analyze how these factors will affect the business.

The macro-environment also has its own significance to be taken into consideration, as it consists of various factors that are not within our control. First of all, the political factor plays the crucial role in every country, as it can affect the business operation in many ways. The strategies adopted by government can remove many obstacles and offer several substantial incentives for foreign investors.

Second, the economic factor is important in determining people's purchasing power. According to the Index Mundi (2010), if a country's purchasing power drastically increases with time, this means that its standard of living is likely to increase, and people will be willing to spend more on the offered goods and services.

Third, regarding the social factor, it is imperative that local firms understand the tastes, needs, and preferences of consumers in the international market. According to the Index Economic Forum (2011), it is important to know the population growth of a certain country in accordance with its growing potential at the same time. All local firms must thoroughly investigate the social environment of the international target market to reduce the risks of failure.

Fourth, technology is a very important and useful tool for expanding the business. For example, the number of internet users in the world has been increasing drastically, as the internet has become a favorable tool to promote a product or service, can deliver the message to the audience very fast and effectively, and moreover, can reduce the cost.

Fifth, the basic facilities and amenities, all means of transportation, communication channels, and security have to be environmentally friendly. This can actually add more value to the foreign businesses, as they can enjoy the benefits of convenience.

9.2.3 Pevvnetration

Penetration here refers to international business expansion and the necessary strategies and tools to be developed to conduct the most suitable platform for simplifying the major hurdles of a local business.

Today, the internet and the latest communication channels help local market capacity to penetrate more easily into the international market, though we aim to further critically analyze the best way of implementing the right and viable marketing strategies to grow most effectively in the targeted international market.

9.3 Six Steps

9.3.1 Orientation and Collect Key Data

The main question of the orientation is "What is our objective?" in this project. This means to orient ourselves and to understand the project, the process, and the people involved. Then, we organize how we propose to plan and schedule the planning. The main issue is "What do we do?" and "How do we do it?" (Muther 2011):

- Understand the project: What? Why? Who? When? Where?
- Understand the purpose or objective(s), the external conditions, the situation(s), the scope/extent, any budget limitations, and the desired form of our planning output
- Understand and document the planning and people issues
- Make a schedule for the project planning

We can use the "Project Orientation Worksheet" that is designed by Muther (2011). Table 9.1 has three components:

- Project essentials

Table 9.1 Project Orientation Worksheet

Description ———————————————————— Project no:——————

Who is responsible? ———————— Authorized/Initiated by ———— Date ————

When project starts ——————— When planning starts ———— Sheet of ——

Project essentials

1. Project objective(s) ————————————————————
2. External condition(s) ————————————————————
3. Situation(s) ————————————————————
4. Scope/extent ————————————————————
5. Form of output ————————————————————

Planning issues	Imp.	Resp.	Proposed resolution	Ok'd by
1.				
2.				
3.				
4.				

Planning schedule																Notes and act
Task or action required to plan	Who															
1.																
2.																
3.																
4.																

Reference notes: ————————————————————

Source: Muther, R., *Planning by Design*, Institute for High Performance Planners, Kansas City, 2011.

- Planning issues
- Planning schedule

In "Project Essentials," enter

- The objective(s) or purpose(s) or goal(s) of this project
- The external conditions, such as synchronization with other projects, specific limitations, overall policies, or larger operational procedures …

- The situation(s): Physical, procedural, and personal situation circumstances
- The scope/extent of the project: How big? How detailed? When needed?
- The form of this planning output: Written report? Action plan approved?

In "Planning Issues," we enter each problem, uncertainty, and question–one line for each—on the left. In the first column, record how important the issue is to this project. Here, we enter a vowel-letter as our order-of-magnitude judgmental rating:

A: Absolutely important
E: Especially important
I: Important issue
O: Ordinary important
U: Unimportant

In the "Responsible" column, enter who is responsible for getting the issue resolved, and mark the initials of the approver.

In "Planning Schedule," list each action required to plan what we intend to do to prepare a market penetration plan. List one action on each line, and show who is responsible for doing it. Set a calendar schedule at the top of the vertical lines.

9.3.2 Clarify Vision

Our first fundamental is to understand the Vision of the company, and whether or not it contains an international expansion ideal.

Before a company expands its business across the world, there are three important systems that must be considered: political, economic, and legal systems. Each country has different types of system that might create problems, obstacles, or opportunities for businesses.

The basic interest for the companies that may want to go abroad is to eventually enjoy the benefits of the strong brand names created in the countries of their origin and to earn higher profits than they can imagine under certain circumstances. However, there are some barriers to entering into the international market, which need to be taken very seriously. These are considered in the next step.

9.3.3 Clarify Environment and Relate to Vision

In this step, understanding the cultures, values, and political economic status of the prospect country or region is essential, as entering into the international market needs to be taken very seriously.

9.3.3.1 Cultures and Values

We need to consider the cultures, traditions, norms, and values of the region where we desire to go global. Therefore, it is important to identify the differences between cultures before penetrating into the market (Sullivan et al. 2008):

- *Power distance:* A higher power distance shows that a society accepts an unequal distribution of power, and people understand where they stand in the system. Besides that, organizations are usually centralized, with strong hierarchies and large gaps between top management and employees. However, low power distance shows that power is shared equally, such as in flatter organizations.
- *Individualism:* This describes the relationship strengths and ties of people within the organization. High individualism shows a loose connection with other workmates because of a lack of interpersonal connection and a share of responsibility. Low individualism shows a society with high collectivism, loyalty, and respect for one another as a group.
- *Masculinity:* This refers to the differences between the male and female roles in the organization or society. This is where dominance occurs, such as when high masculinity indicates that man are masculine and woman are feminine, men are expected to be strong and tough, and there is a distinction between men's and women's jobs. Low masculinity means that women and men are treated equally.
- *Uncertainty avoidance:* High uncertainty avoidance shows that people in the organization avoid uncertain situations whenever possible because they need defined structures in which differences are avoided. In contrast, low uncertainty avoidance shows that organizational values differ, and organizations prefer informal business attitudes.
- *Long-term orientation:* High long-term orientation has positive effects on strong work ethics, high importance of training

employees, and so on. Low long-term orientation shows promotion of equality and self-actualization.

9.3.3.2 Political Economy   This can also be defined as the relationships of a country as a whole. In general, it is the study of economics, politics, and legal systems with the purpose of explaining the existing relations among different countries, since each country has different political economy.

In the modern world of globalization, political economy is not only associated with politics, economics, and legal systems; it also includes technology, international relations, human psychology, and many more. Lastly, political economy has always been one of the core tools used to analyze the trends of the current global economy.

9.3.3.2.1 Political System   A political system is a form of government that consists of a group of people who have the authority to decide how to manage their nation. It is a social system of politics and government that is interrelated with the legal and economy systems. According to Easton (1965), the political system is the authoritative allocation of values in society. It consists of political organizations, institutions, interest groups such as political parties and trade unions and the relationship between those institutions, political norms, and rules and regulations that administrate the nations.

9.3.3.2.2 Economic System   The economic system is the way a society organizes the production, consumption, and distribution of goods and services. It is composed of people, institutions, and their relationships. It deals with the problems that occur in the economy, such as allocation and scarcity of resources. It focuses on what and how much will be produced, how it will be produced, and for whom it will be produced.

9.3.3.2.3 Legal System   The legal system consists of rules and regulations or laws that regulate the behavior of business practices. It classifies how companies, either domestic or private, carry out their business transactions. A domestic firm must follow the laws and customs enforced by its country, and an international firm must abide by the law of both its home country and the host countries in which it operates. For instance, China used to enforce laws that prohibited foreign investment and restricted foreign trade. Then, laws were passed permitting joint ventures using Chinese firms and helping China to increase its market potential, market performance, infrastructure, and strategic position.

It is important for an international business that operates its business in a different country to abide by the laws of that country. Different courts in different countries could make judicial decisions that affect day-to-day business transactions, such as embargo and extraterritoriality. It is important for an international business involved in shipping products to another country to obey with the rules and regulations of that country (Beatty et al. 2015):

- *Common law:* Originally British, this is based on the concept of precedence, custom, and tradition. The facts and decisions of a particular case are determined based on previous cases that have had similar facts and situations rather than written laws. Under this law, courts play an important role in interpreting the law. Nations that practice common law include Pakistan, Hong Kong, the United Kingdom, and Barbados. For instance, manufacturers of faulty products are more vulnerable to lawsuits in the United States than in the United Kingdom as a result of evolutionary differences in both countries' case laws.
- *Civil law:* This focuses on resolving noncriminal activities such as disagreements over contracts, property ownership, child custody, property damage, and divorce. Civil law is codified and not determined by judges based on precedent. It has to be written and published before it can apply to people. It consists of various types of civil cases, such as consumer law, international law, employment law, and business law.
- *Theocratic law:* Also known as *religious law*, this is based on religious doctrine, precepts, and beliefs. It is a form of government in which God is recognized as the state's supreme civil ruler, because it carries out the interpretation of the will of God as set out by religious scripture and authorities. Laissez-faire government attitudes toward trade and investment policies remove the barriers to internationalizing trade

and enable firms to view the world as their market. Besides this, government increase in foreign acquires of corporation, reduce tariff barriers allowing local consumer to have higher buying power and increase in level of world trade.

9.3.4 Develop Penetration Plans

Local companies need to find their critical competitive advantages in foreign markets to penetrate successfully. It is vital to take into consideration the political, economic, and legal issues of the targeted foreign market, and moreover, the culture, norms, and values and the acceptability of the offered products or services play an important role in penetration into the international market.

The most viable strategy for a local to go global is to start its business via joint venturing. In this way, the local companies in foreign countries will understand the legal, economic, and political situations better than foreign businesses. The franchising model can also play a very important role, as this model involves fewer risks compared with others. The franchisor bears less involvement and responsibility for its business going international.

By considering these two fundamentals (understanding the scope and the factors of globalization), we can get a clearer idea of what essential measures are needed for local firms to prosper in the foreign market.

In this step, considering these fundamentals, we need to develop a penetration plan for the targeted market via marketing, technological, promotional, and other means (Hill 2007):

- *Technology:* An efficient and productive business always ensures that its machinery, technology, and information are up to date. Technology changes the work the business does and the way business operates, such as the relationships between suppliers, producers, retailers, and customers. Besides this, technology differences drive firms to plan their products and their sales on a global basis. Airplanes, televisions, telephones, and computers allow information flow from one place to another without direct meetings.

  Moreover, communication, information-processing, and transportation technologies have been dramatically improved by technological advancement. Cell phones, emails, the internet, and microprocessors make it easier for firms to communicate and exchange information with each other, which in turn, leads to efficient production. The growth of the internet has created a low-cost global platform for communicating and doing business among people across borders. For example, previously, to book a flight ticket we had to travel to a retail outlet; nowadays, it can be done by a click of the mouse.

- *Market:* The demand for goods and services will increase, which leads global customers to the companies expanding globally. If a firm's customers are other global businesses, globalization may require it to reach these customers in all their markets. Furthermore, global customers require globally standardized products. Common market needs and global marketing channels make globalization easier. If firms only operate their businesses locally, the income that they earn will be limited. Besides, goods will not be upgraded due to the lack of exposure.

- *Competition:* Competition is a contest among individuals or groups. It is the concept of firms striving for a greater market share. Competition continues to increase dramatically due to the increase in the number of manufacturers and service providers, which may result in more countries becoming key competitive battlegrounds. Usually, global competitors have cost advantages over local firms.

9.3.5 Evaluate and Accept the Best Plan

Alternative penetration plans should be separately evaluated, and the best applicable one should be selected using the Evaluation of Alternatives Worksheet. Identify each alternative appropriately, as described in Section 9.3.4. Prepare a worksheet showing the comparable costs of each alternative. Also, on a separate worksheet, make a comparison of the intangible benefits and risks of each alternative. Compare alternatives based on costs, benefits, and risks, select the best, and get the others to approve it (Muther 2011).

Enter the headings on a fresh copy of the worksheet (see Table 9.2 generated by Muther (2011), checking the box marked *intangibles*

Table 9.2   Evaluation of Alternatives Worksheet

Evaluating alternatives		Project _____ Number _____
[ ] Costs: Estimated by .... Approved by		By _____ With _____
		Date _____ Sheet _____ of ____
[ ] Intangibles: Weight set by .... Tally by		Description of alternatives:
Ratings by .... Approved by		X.
Evaluating description		Y.
A = Almost perfect, O = Ordinary result		Z.
E = Especially good, U = Unimportant results		V.
I = Important result, X = Not acceptable		W.

Factor/consideration		WT.	Alternative				
			X	Y	Z	V	W
1.							
2.							
3.							
4.							
5.							
6.							
7.							
8.							
9.							
10.							
11.							
12.							
13.							
14.							
15.							
Total	Annualized cost (line____ plus line____)						
	Weighted rated down total						

Reference notes:

a. _____   d. _____
b. _____   e. _____
c. _____   f. _____

Source: Muther, R., *Planning by Design*, Institute for High Performance Planners, Kansas City, 2011.

(upper left). Identify each alternative by a letter—X, Y, Z—and give a brief two-to-five-word description of each.

List all factors, considerations, or objectives the organization wants the project's intended plan to achieve. Select, or ask your approvers to select, the most important factors. Then, ask them to weigh the importance of each factor relative to the most important (10). Indicate each selected weight on the worksheet, and record by whom the weight values were determined.

Ask your operations team and/or staff members who will use the proposed plan when installed to rate, for each factor, the effectiveness of each alternative in achieving that factor. Use A, E, I, O, or U to represent the descending order of effectiveness, as noted in the upper left-hand box of the worksheet. Enter, in the small rectangular *boxes within boxes* on the form, the selected vowel-letter ratings. Record the name(s) of the person(s) doing the rating.

After rating all alternatives for each factor, convert letters to numbers (A = 4, E = 3, I = 2, O = 1, U = 0) and multiply the rated number by the respective weight value. Enter the resulting weighted-rated values on the worksheet.

Down-total the weighted-rated values for each alternative, enter into the worksheet, and record by whom the tally was made. The alternative with the highest total should be the *winner*—subject to cost factors determined separately. In the lower left corner, indicate that these are weighted-rated down totals. Record any explanatory notations at the bottom, suitably referenced by an encircled lower-case letter.

9.3.6 *Implementation Plan*

This step is dedicated to carrying out the selected plan. In implementation, who does what is very important. It includes the actions needed to make the plan come true. Who will be responsible and the duration of action need to be clarified. This step sets the framework for dealing with the *expectations* on time. We can use the MS Project tool in this step.

References

Atlas, R. I. 2013. *21st Century Security and CPTED*, 2nd ed. Boca Raton, FL: CRC Press.

Beatty, J. F., Samuelson, S. S. 2015. *Business Law and the Legal Environment*, Std. ed., 7th ed. Boston, MA: Cengage Learning.

Drucker, P. 2002. *Innovation and Entrepreneurship*. New York, NY: HarperCollins.

Easton, D. 1965. *A Systems Analysis of Political Life*. New York, NY: Wiley.

Hill, C. W. L. 2007. *Global Business Today*, 5th ed. Columbus, OH: McGraw-Hill.

Jaman, M. 2012. Critical analysis of segmentation strategy for potential product launch. *International Journal of Scientific & Technology Research*, 1(11): 62–65.

Jerrard, B. 2004. *Managing New Product Innovation*. London, UK: Taylor & Francis.

Johnson, R. 2013. *Antiterrorism and Threat Response*. New York, NY: CRC Press.

Kim, C., Mauborgne, R. 2005. *Blue Ocean Strategy*. Boston, MA: Harvard Business School Publishing.

Kotler, P. 2008. *The New Strategic Brand Management*. London, UK: Kogan.

Kotler, P. 2010. *Ingredient Branding: Making the Invisible Visible*. London, UK: Springer.

Kotler, P., Armstrong, G. 2011. *Principles of Marketing*, 14th ed. Harlow, England: Pearson Hall.

Kotler, P., Keller, K. 2012. *Marketing Management*, 14th ed. Upper Saddle River, NJ: Prentice Hall.

Kotler, P., Lee, N. 2005. *Corporate Social Responsibility*. Toronto: Wiley.

Lannuzzi, A. 2012. *Greener Products*. New York, NY: CRC Press.

Mahan, S., Griset, P. L. 2013. *Terrorism in Perspective*, 3rd ed. London, UK: Sage.

Muther, R. 1988. *High Performance Planning*. Kansas City, MO: Management and Industrial Research Publications.

Muther, R. 2011. *Planning by Design*. Kansas City, MO: Institute for High Performance Planners.

Naquin, S. S., Holton, E. F. 2006. Leadership and managerial competency models: A simplified process and resulting model. *Advances in Developing Human Resources*, 8(2): 144–165.

O'Sullivan, D., Dooley, L. 2009. *Applying Innovation*. Thousand Oaks, CA: Sage.

Overton, R. 2007. *Feasibility Studies Made Simple*. Sydney, Australia: Martin Books.

Porter, M. E. 1998. *Competitive Strategy: Techniques for Analyzing Industries and Competitors*. Florence, Italy: Free Press.

Renvoise, P. 2007. *Neuromarketing*. Nashville, TN: Thomas Nelson.

Ries, A. 2002. *The Fall of Advertising and the Rise of PR*. New York, NY: Perfect Bond.

Samli, C. 2011. *From Imagination to Innovation*. New York, NY: Springer.

Sandberg, B. 2008. *Managing and Marketing Radical Innovations*. London, UK: Routledge.

Sportack, M. A., Pappas, C. F., Rensing, E., Konkle, J., Smith, R. C., Welk, D., Short, D. 1997. *High Performance Networking*. Indianapolis, IN: Sams Publishing.

Sullivan, D. P., Daniels, J. D., Hradebaugh, L. 2008. *International Business*. London, UK: Pearson.

Trail, B. 1997. *Products and Process Innovations in the Food Industry*. London, UK: Chapman & Hall.

Trott, P. 2005. *Innovation Management and New Product Development*. London, UK: Pearson Education.

Wen, Y. 2011. *Networking Enterprise IP LAN/WAN Design*. San Jose, CA: The System Administration Company.

Wheat, B. 2003. *Leaning into Six Sigma*. New York, NY: McGraw Hill.

Index

Printed and bound by CPI Group (UK) Ltd, Croydon, CR0 4YY

28/10/2024

01780264-0002